D1785695

SAVING AUSTRALIA

SAVING AUSTRALIA

A BLUEPRINT FOR OUR SURVIVAL

Vincent Serventy

CHILD & ASSOCIATES
AN ALL-AUSTRALIAN PUBLISHER

Front cover photograph: Dense rainforest hangs over a waterfall in the Lamington National Park.

Published by Child & Associates Publishing Pty Ltd,
5 Skyline Place, Frenchs Forest, NSW, Australia, 2086
A wholly owned Australian publishing company
This book has been edited, typeset and designed
in Australia by the Publisher
First edition 1988
Reprinted 1990
Text and photographs by Vincent Serventy
© Vincent Serventy and Associates 1988
Edited by Sheena Coupe and James Young
Printed in Singapore by Kyodo-Shing Loong Printing Industries Pte Ltd
Typesetting processed by Deblaere Typesetting Pty Ltd

**National Library of Australia
Cataloguing-in-Publication Data**

Serventy, Vincent, 1925–
    Saving Australia.

    Bibliography.
    Includes index.
    ISBN 0 86777 157 7.

    1. Environmental protection—Australia.
    2. Conservation of natural resources—
    Australia. I. Title.

363.7'00994

All rights reserved. No part of this publication may be reproduced, stored in a retrieval system, or transmitted in any form or by any means, electronic, mechanical, photocopying, recording, or otherwise without the prior permission in writing of the publisher.

Contents

Acknowledgements

It would be impossible to acknowledge all the people who have helped me with this book but Dr George Wilson in particular was of great assistance in the early stages. The following organisations helped to provide the facts I needed.

Firstly, the Australian Heritage Commission. My fellow commissioners and the staff helped a great deal, including reading drafts of particular chapters. The commission, one of the 'lean and hungry' organisations to be found in all civil services, is rarely given due credit for its dedicated work.

The Australian National Parks and Wildlife Service has always been of immense assistance. Both these groups are under the umbrella of the Department of Arts, Sport, the Environment, Tourism and Territories, which carried out the organisation of the National Conservation Strategy and has tried to keep alive interest in this grand concept. Without its publications and other assistance this book would not have been possible. I have served on some of the committees and have always enjoyed the experience.

The department has also encouraged a much needed unity among the conservation groups around Australia, both government and non-government. When critics talk of 'economising' on such public servants they show they have no understanding of how great the loss would be to all Australians if these groups were unable to carry out their work.

The state services complement the federal ones. They are organised under a diversity of titles but always incorporating words such as wildlife, conservation or environment. They have always been willing to assist in anything that contributes to the grand conservation design and are listed in every state capital telephone book.

Above all, there are the hundreds of thousands of members of non-government conservation and natural history societies that I have met in the last 40 years. Their publications are essential to gain a balanced view, and on this broad base of dedicated individuals, conservation has become accepted in official and unofficial quarters.

As individuals these enthusiasts get little praise and often much abuse, both verbal and even, at times, physical. They are the real heroes and heroines in saving Australia. Among them are those working on environmental education. They ensure that recruits from among our children will fill the gaps created by old age. All of us have accepted the dictum that we are not only inheritors of our past but, more importantly, also trustees for the future.

Without our wholehearted acceptance of the principles and practice of the World Conservation Strategy and the National Conservation Strategy for Australia, that future will be in doubt.

A personal statement

Australia in 1988 celebrated a bicentennial of settlement. Not of the first human arrivals, as that was possibly as long ago as 120 000 years, probably about 70 000 years ago, and certainly at least 40 000 years ago. How many waves of humans poured into this great south land is at best an informed guess. Perhaps there were three major invasions, each overwhelming or absorbing the earlier hunter-gatherers.

In 1788 came the Europeans, agricultural people armed with a superior technology, both for peace and war. In the face of this onslaught, the Australian Aborigines melted away, mainly because their country was taken. It is a truism of nature conservation that the quickest way to destroy a species is to destroy its habitat. That is as true for humans as it is for koalas or kangaroos. Yet the Aborigines had shaped the land to their own needs; what the white settlers called 'natural' was a human artifact. The change was effected by the use of fire, mainly for hunting.

The new people possibly brought the dingo into this giant Southern Ark. Those most intelligent and largest of predators, humans, probably pushed into extinction the giant kangaroos, marsupial lions and diprotodons, already hard pressed by climatic changes. This disappearance of giants—mammals, birds and reptiles—took place all over the world and coincided with the arrival of humans in all continents except Antarctica.

The Europeans came with new predators and new methods of using the land. Massive changes took place, not all good, not all bad. The trickle of extinction became a flood. Soils began to degrade. Minerals rose to the surface due to the destruction of the forests and these areas became salty wastelands. Yet new crops and farm animals raised food production to high levels, sustaining far more people than could be nourished by hunting animals and gathering plants.

Now is the time to take stock of what has happened and to learn from the past so that we can plan for the next 200 years of a new Australia. It will be a new kind of nation where original Australians and new Australians coming from all parts of the globe can work in harmony.

Vincent Serventy

1. Does Australia need saving?

Here are some facts gleaned from recent reports on Australia's current position:

- A joint commonwealth–state study on soil conservation carried out between 1975 and 1977 found that 90 per cent of the agricultural land in New South Wales alone needed some type of soil conservation treatment. In the non-arid zone between 16 and 50 per cent needed improved land management and between 30 and 45 per cent required soil conservation works. Not a comforting thought after only 200 years of European settlement. What applies to this state applies to all the others.
- Soil salinity is on the increase in Australia. About 500 000 hectares of dryland agricultural soils and between 85 000 and 120 000 hectares of irrigated lands are affected

Land degradation: proportion of agricultural land needing treatment, states and territories, 1975

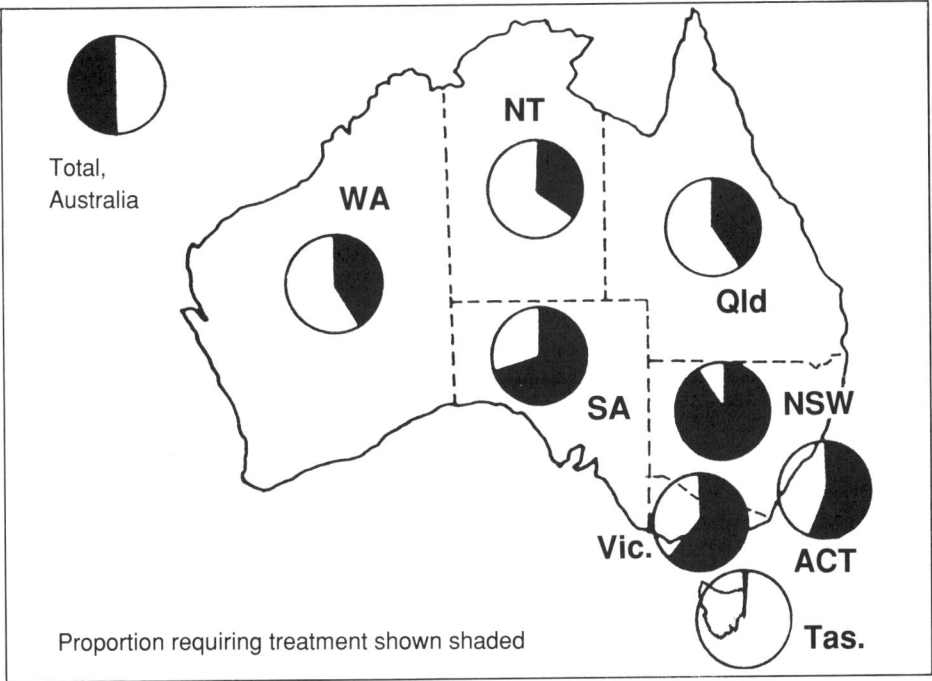

Source: Prepared from data in Woods, L. E., *Land Degradation In Australia,* AGPS, Canberra, 1983, and Department of Arts, Heritage and Environment.

by soil salinity. This is a problem for all states with south-western Australia and the Murray River system being the most seriously affected.

- Since European settlement two-thirds of our forest areas have been cleared and the remaining forests cover only about 5 per cent of our total land area.
- Our rainforests are the home of about half our vertebrate animal species, half of our plant species and a third of our invertebrate animals. Today rainforest occupies 0.25 per cent of our land area, having been reduced by three-quarters since European settlement.
- Since European settlement 15 species of native mammals and one bird species are probably extinct. Rare and endangered are 26 mammals, 22 birds, 9 reptiles and 6 amphibians. The position of our fish and invertebrate animals is almost unknown. At least 2206 plant species are rare or threatened.

Widespread salinisation of streams in the south-west due to clearing of native vegetation in catchments has caused ecological damage and led to a loss of 36 per cent of the estimated original fresh-water resources. Progress is being made in stabilising selected catchments, but no significant advances have been made in rehabilitating seriously degraded catchments.

(A State Conservation Strategy for Western Australia, 1987)

This tale of disaster could go on and on. Some of the dangers facing Australia are discussed in detail in the chapters to follow. There are many others. Air pollution in many of our cities reaches levels dangerous to health. Oil spills damage rivers, estuaries, beaches and oceans. The crown-of-thorns seastar is ravaging a large section of the Great Barrier Reef. Inedible shrubs are spreading over many parts of arid Australia.

Farm factories are already with us. Intensive cropping is creating agricultural regions with as much visual and wildlife impact as a barren plain. Battery farming for hens, pigs, calves and other domestic stock offends every sensitive person. Clearfelling is changing our forests into even-aged stands of single species with only slightly more interest than a field of wheat. In the sea, dredging cuts a swathe through the whole marine resource, using only one species and leaving the rest to die.

Many of these changes have been so gradual that we are becoming adapted to a poorer quality lifestyle in our cities and our natural heritage is slipping into degradation without our even noticing.

It is salutary to stand on North Head at the entrance to Sydney Harbour and look towards the city. From there you can see the blanket of brown filth that three million people breathe, day in, day out. History has shown that humans can adjust to the most frightful situations but is there any reason why we should accept such conditions when the remedy is in our own hands?

Caring means planning for a better future. As William Shakespeare wrote in *Julius Caesar:* 'The fault, dear Brutus, is not in our stars, but in ourselves.' If we slide down the easy path to ecological disaster with the assumption that it can't be helped, it's 'progress', the fault is ours. What most of us call progress is only change—and not necessarily change

The total area of standing rainforest (all types) in Australia, represented schematically as two dots

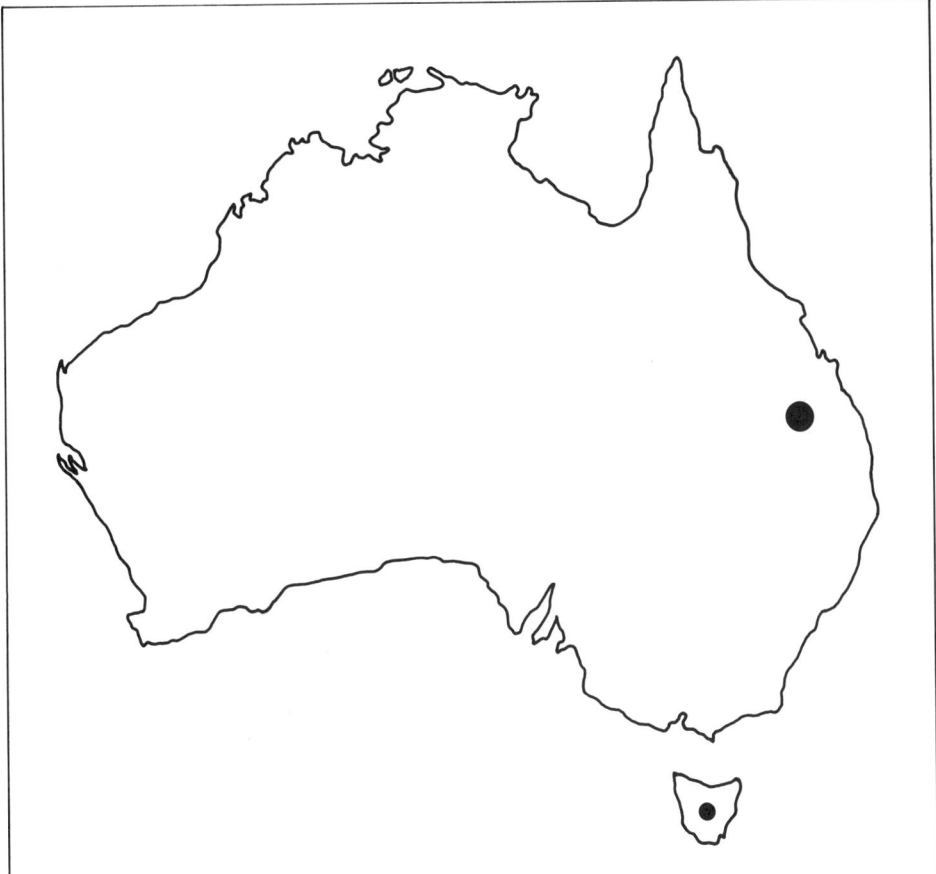

The areas for the mainland and Tasmania are 15 680 square kilometres and 4561 square kilometres respectively and the radii of circles of these areas are 70.6 kilometres and 38.1 kilometres.

Source: World Wildlife Fund.

for the better. The late Judge Barry expressed the most pessimistic view when he envisaged a future Australia as 'an empty quarry surrounded by an oil slick.' This seems unlikely today but another, more subtle, danger was stressed by Sir Macfarlane Burnet in a Boyer Lecture when he said: 'Who would want to live in an overpopulated world, where outside the urban sprawls there are only vast agricultural factories in the fields?'

Surely we have an obligation to leave for future generations the present diversity of landscapes, the wonder of our animals and plants and all the other attractions we enjoy today? We are not only inheritors of this land but trustees, with a duty to hand on our trust in at least as good a condition as it is at present. We should strive to hand it over in an even better condition, one with the mistakes of the past corrected. This is what a conservation strategy is all about.

This book is dedicated to the following statements:

World Conservation Strategy Document 1980

Ultimately the behaviour of entire societies towards the biosphere must be transformed if the achievement of conservation objectives is to be assured. A new ethic, embracing plants and animals as well as people, is required for human societies to live in harmony with the natural world on which they depend for survival and wellbeing. The long-term task of environmental education is to foster or reinforce attitudes and behaviour compatible with this new ethic.

National Conservation Strategy for Australia (ncsa) 1983

The three main objectives of living resource conservation identified with the World Conservation Strategy have been adopted for the National Conservation Strategy for Australia (NCSA).

They are:

(a) to maintain essential ecological processes and life-support systems (such as soil regeneration and protection, the recycling of nutrients, and the cleansing of waters), on which human survival and development depend;

(b) to preserve genetic diversity (the range of genetic material found in the world's organisms), on which depend the breeding programme necessary for the protection and improvement of cultivated plants and domestic animals, as well as much scientific advance, technical innovation, and the security of the many industries that use living resources;

(c) to ensure the sustainable utilisation of species and ecosystems (notably fish and other wildlife, forests and grazing land), which support millions of rural communities as well as major industries.

An additional and no less important objective is:

(d) to maintain and enhance environmental qualities which make the earth a pleasant place to live in and which meet aesthetic and recreational needs.

These statements are written in the jargon that is inevitable in such documents. In everyday language they say first and most importantly that we need a new ethic, a sense of values, to guide our future actions. We know that spaceship earth is a place of finite resources. We know, too, that it will be our only home in the foreseeable future. Dreams of moving to new planets are science fiction and will not become fact in time to save us from our folly.

We must learn that we are part of nature, not outside it, or even more important, above it. The environment is more fragile than we once thought and we must learn to live in harmony with nature, rather than try to subdue it to our short-term greed. The concept of trusteeship sums up this new ethic. If we accept it, not only will we be happier but, paradoxically, we will live more richly when we learn nature's rules.

Using this new ethic we can look at the four objectives of the NCSA and how they should guide our environmental lives:

(a) means that all life on earth and the resources that sustain it are a web of interlocking devices. Like a spiderweb, if we poke into it with too blunt a finger, the whole structure may come tumbling down. We have seen this happen in history with the collapse of the great irrigation civilisations along the Tigris and Euphrates rivers, as well as other cultures whose relics remain today as a warning;

(b) states that in our plants and animals we have a treasurehouse of diversity. This is what scientists call a gene pool, since the genes in life cells carry the hereditary characters. Into this vast pool we may dip forever to obtain our needs. Perhaps a humble rainforest plant may one day provide a cure for cancer. Even today almost half the medicines we use are plant products. The grasses in the mountains may give new vigour to a declining crop plant. If we allow this gene pool to become impoverished due to stupidity or ignorance, which is a kind of stupidity, and greed, which is a greater stupidity, we will suffer in the future;

(c) states that we must not kill the goose that lays the golden eggs. History is full of examples of such killing. Vast herds of bison were diminished because of greed. Passenger pigeons in their hundreds of millions disappeared under the hunting onslaught. The great auk and dodo were two among the many island creatures that became extinct. We hunted the giant whales almost to extinction. In Australia we have destroyed grasslands by overgrazing and salted fertile soils by unwise tree clearing;

(d) is interesting because it was added to the NCSA by Australians. We hope that when the World Conservation Strategy is updated in the near future our suggestion will become universal. Humans do not live by bread alone!

If Uluru and the Olgas were levelled because their rock became economically valuable, Australians would survive, but we would be spiritually impoverished. If the world's tallest hardwood trees, the mountain ash of the Styx forest in Tasmania, were felled to make woodchips, in 300 or 400 years others could grow. Yet for centuries we would no longer enjoy the excitement and sense of awe we have in looking at those forest giants.

For some this is not sufficient argument; these are people who rarely lift their snouts from the trough of life to look at the glory of the stars. Yet the community may suffer. History warns that such cultures are less efficient and are overtaken by those which keep diversity both in natural and human resources.

History of the strategies

'God knows we need some good environmental commonsense in this beautiful, fragile, oft desecrated land of ours. . .' wrote Professor W. D. Williams, Dean of the Faculty of Science in the University of Adelaide, an authority on our freshwater life and a hardworking conservationist. He was commenting on the need for a more responsible and responsive attitude from both people and governments.

Williams is only one of the thousands of scientists around the world who care for our fragile spaceship earth. Together with naturalists, conservationists, public servants and many others, they decided to do something to help achieve environmental commonsense. This is what the strategies are all about.

World Conservation Strategy

In 1980 the World Conservation Strategy was launched in some 30 countries around the world. It was developed by the International Union for Conservation of Nature and Natural Resources (IUCN), the United Nations Environment Programme (UNEP) and the World Wildlife Fund (WWF).

In Australia, the then Prime Minister, Malcolm Fraser, accepted the strategy on behalf of the Australian government. All the states took similar action, indicating that here was a document above party political divisions. I had the privilege of launching the strategy in New South Wales and my enthusiasm for its potential has increased steadily.

The World Conservation Strategy is a global one but it can succeed only if every nation develops its own strategy for dealing with problems inside its borders, as well as accepting responsibility for environmental problems outside its national boundaries but not outside its international interests.

We must share the responsibility, as well as enjoy the benefits, of the oceans and the atmosphere. Also, as has been shown so clearly with the theory of the nuclear winter that will follow an atomic war, we all float or sink together where major environmental problems are concerned. Since 1980 some 32 nations have either completed or are working on their own national strategies.

National Conservation Strategy for Australia

In 1980 the Australian government set up a national steering committee with representatives from the commonwealth, states and the Northern Territory. A task force from the Department of Art, Heritage and Environment carried the work forward. In December 1981 a national seminar considered a number of documents on living resource conservation. From this meeting came a further discussion paper, *Towards a National Conservation Strategy,* that was given a wide circulation.

Also from this seminar came a consultative group of four nominees each from industry and conservation groups who met the steering committee to carry on the work. I was fortunate enough to be one of them. By the end of 1982, 550 written submissions on the discussion paper had been received.

Using this material the task force prepared a draft strategy for discussion at a national conference. A federal election caused a postponement but the meeting was held from 10 to 13 July 1983 and the 153 delegates represented every facet of interested opinion. It was an exciting four days. Debate in the general sessions and workshops was loud and long. The few delegates who tried to espouse their own specialist views were soon swept along with the tide of popular opinion. A belief that we were developing the most important document in the history of Australia helped us reach a consensus.

Chairman Sir Rupert Myers summed up the general feeling when he wrote to the Minister of Arts, Heritage and Environment that 'the document is one of great significance; Australia can be proud of it'. He also summed up our feelings when he wrote:

> The development of a National Conservation Strategy for Australia is a significant step towards the maintenance and sustainable use of our living resources. Just as wide ranging

community contributions assisted its development, so widespread community endorse-
ment can set the Strategy into place as a blueprint for further action. Ultimately the
Strategy can only succeed if its objectives are accepted and incorporated as a fundamental part of the national ethos.

The four main objectives of the strategy were listed earlier in this chapter; the entire
document is reproduced in Appendix 2. Present in our minds at the conference was the
belief that overshadowing all our deliberations were the new 'horsemen of the apoca-
lypse'—exploding human populations, the profligate use of the earth's non-renewable
resources, and nuclear war. We must play our part as a nation in helping rein in those riders
taking us at full gallop to global disaster.

The document pointed out the urgent need to educate the public in the importance of
environmental planning, not only for the next few years but for our long-term future. It also
recommended that an interim consultative committee be formed to involve government,
industry, research, education and community interests. The government accepted this
advice and I was a member of the committee. We have done our work and sent a report for
further action. Our suggestions, and how well both the Australian and state governments are
carrying out our hopes, are discussed later in the book.

We now have a Magna Carta for conservation. The road ahead has been defined and
whether we walk along it is up to us.

2. Ecological realities

'Dr Who' and other science fiction programmes, as well as 'soapies' dealing with doctors and nurses, have made the term 'life support system' well understood. What is true for a human in a spaceship or an intensive care ward also applies to the spaceship earth.

We all know that life depends on energy from the sun. Plants use sunlight to provide the food all creatures need. When any plant or animal dies, a mass of fungi, bacteria and other lifeforms break it up into material that can be reused. This has been happening since life began on earth, yet too few of us realise that unless we keep the basic foundation stones of soil, water and air in good shape, we face disaster.

Ecology is a technical word which has earned common usage during the last few decades. It means a study of homes, and the NCSA is all about a study of our Australian home and a plan to live in it with increasing comfort for the next 200 years.

Popular opinion, backed by optimism and a diet of science fiction in books and films, is convinced that a technological solution is possible for any problem. Any solution, however, must work within the bounds of ecological realities. These cannot be swept away by any human invention and we ignore them at our peril. We must play the conservation game by nature's rules. She is a fair opponent and will richly reward the player willing to learn them. However, she will checkmate those who think they can escape the rules, without passion but with certain extinction.

The World Conservation Strategy document highlights the urgency of living resource conservation. The warning was reinforced by the findings of the Global 2000 study prepared by the United States Council on Environmental Quality and the United States Department of State. Both warned that if present trends continue, by the year 2000 the world will be less stable ecologically, deserts will increase, water shortages already grave will become worse, the world's forests will continue to shrink, the soils on which agriculture depends will deteriorate, more plant and animal species will become extinct, carbon dioxide levels will rise as will the levels of chemicals which deplete the protective ozone layer, and hazardous chemicals (including radioactive ones) will increase.

The study concludes that the earth 'is imperilled unless there is a keener awareness of current trends, and changes are induced that will alter the projected outcome...'

Soils

Australia is not a lucky country in terms of its soils. Over aeons of time they have been leached while the soils of the higher rainfall areas of the north are generally infertile.

> *They call her a young country, but they lie;*
> *She is the last of lands, the emptiest...*

So wrote A. D. Hope and the poet echoed what geologists have told us from the earliest days of settlement: Australia is an old continent; so too are its soils, and essential nutrients for

Lake Argyle, the huge expanse of water created by damming the Ord River. So far economic returns have been poor with most income coming from tourists who visit this inland lake in northern Western Australia. (See also pp. 26–9)

A koala mother and baby. Millions of these marsupials were killed for their skins about 100 years ago. Public opposition spearheaded by the Wildlife Preservation Society of Australia forced governments to ban their hunting. Today they are regarded as a tourist asset worth many millions of dollars. (See also pp. 59–62)

All turtle species are protected in Australia. Here a female green turtle heads for the sea after spending several hours making a nest for her eggs on Heron Island on the Great Barrier Reef. (See also pp. 59–75)

plant growth are scanty. There are a few regions of good soils such as the Darling Downs and limited areas in Victoria, New South Wales and elsewhere.

Given a poor birthright, the European settlers reduced the continent's potential even further. Today, half the agricultural land outside the arid areas needs treatment for soil erosion. This is a frightening figure. How did it happen? Soil erosion began with the coming of the first people and their intensive use of fire, originally as a hunting tool and later as an embryonic agricultural tool. Anthropologist Dr Rhys Jones called this 'firestick farming'.

Continual burning changed local plant life, favouring grasses that provided seeds for grazing animals such as kangaroos and wallabies. Then came the white settlers. They brought not only a more intense agriculture but a host of new animals and plants. The soils that took so long to be created began to move (the CSIRO has calculated that the agricultural soils of eastern Australia took 30 000 years to form, and in human terms this means there is no new soil being created to replace any losses).

Taking the soil erosion story chronologically, the major impact in those first years was the arrival of hard-hoofed stock with different grazing patterns to a soil which had felt only the gentler feet of our marsupials. An international study has shown that Australians have degraded 11.2 hectares for every person. We are in the unenviable position of having the worst national figures, eight hectares per capita higher than the United States.

Fencing the country to confine stock animals meant that overgrazing became common. Owners made the mistake of watching the condition of their animals instead of watching the condition of the plants. Often the pasture plants were grazed too heavily and with the arrival of dry times, population crashes occurred. In the western division of New South Wales the great drought of 1901-02 brought sheep numbers down from 13.5 to three million. Since then sheep numbers have varied between two and five million, but have never reached the old totals. More recently, a similar slump occurred around Alice Springs. Cattle numbers at a peak of 350 000 in 1969 fell to 130 000 during the drought that followed. South Australia and Western Australia had similar stock crashes.

Major land uses such as grazing, agriculture, urbanisation and industrial development are diminishing species diversity throughout the state; in some localities essential ecological processes continue to be disrupted. In the wheatbelt approximately 60 per cent of the species of medium-sized mammals have disappeared since European settlement. From more than 7000 native vascular plant species recorded in Western Australia, 1024 species are listed as rare or threatened, 83 per cent of these being from the south-west. It is unclear whether the rate of extinction has changed: for some groups it may have decreased in the last 20 to 30 years. However, in most groups of plants and animals, the numbers of some rare species are still declining and discrete populations are still being lost.

(A State Conservation Strategy for Western Australia, 1987)

Even the removal of all domestic stock may not restore the old conditions. The semi-arid woodlands of eastern Australia, covering 500 000 square kilometres, was once a pleasant park-like landscape with plentiful perennial native grasses and clumps of trees and shrubs.

Level of disturbance to Australian plants resulting from European settlement

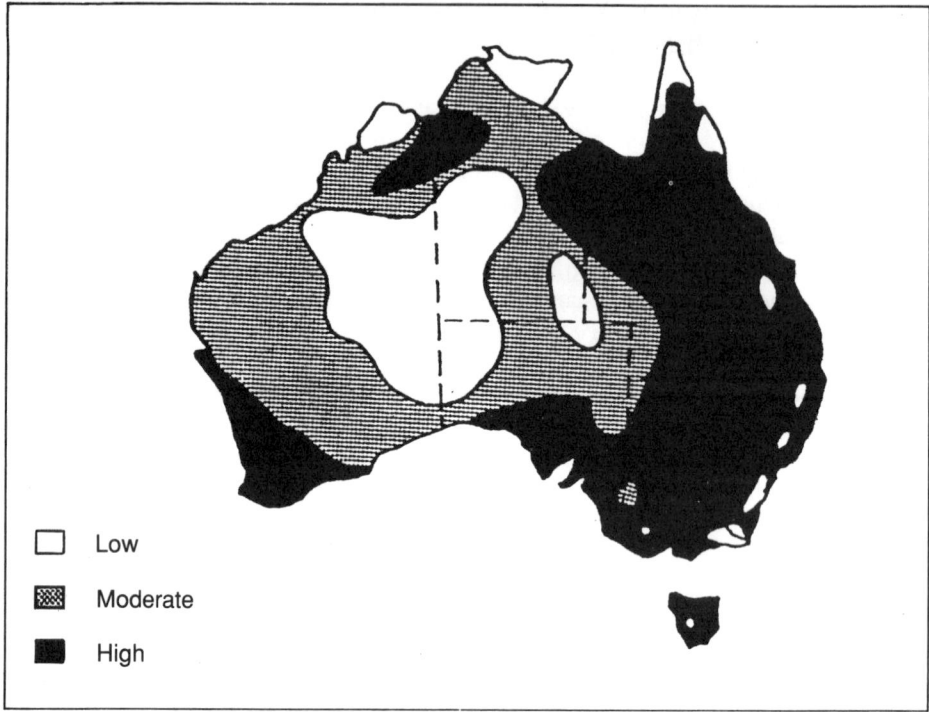

Low

Moderate

High

Source: Adamson and Fox, 1982, in Leigh, J., et al., *Extinct Plants of Australia,* Macmillan, Melbourne, 1984, and Department of Arts, Heritage and Environment.

Overstocking followed by drought changed this. The old pastures did not return and there has been a steady increase of shrubs useless for stock. These 'woody weeds' have created an alarming problem not only in Australia but throughout the world in similar climatic zones. In all cases the pattern of cause and effect has been the same. CSIRO research has shown that in this shrubby country the spread of the woody weeds can be rapid. One paddock near Cobar carrying 5800 shrubs per hectare in 1974 had a 45 per cent increase by 1977.

With good conditions the amount of feed grass can reach 1500 kilograms per hectare; when woody weeds take over, it falls to 200 kilograms per hectare. This is the heart of the problem. Without a fuel level of grass of 900 kilograms per hectare the country cannot carry a fire—and carrying a fire is the best way to get rid of the shrubs. Overgrazing removed the grasses while fencing with watering points kept the stock in restricted areas. Sheep will walk about 5, and cattle up to 13 kilometres to a watering point, even less in saltbush country.

Following the native animals' grazing patterns might offer hope in the future use of our grazing lands. Research by the CSIRO indicates that only management burning will halt the advance of the shrubs.

What is the future of these arid rangelands? Some extreme conservationists have suggested that white settlers should move out and allow the country to revert to one huge national park. Yet even when all the domestic stock is removed there is no certainty the good old days will return. One study at Koonamore station in South Australia found no appreciable improvement over many decades, although at Broken Hill fencing zones from stock has been more successful.

If sheep and cattle were removed, rabbits could destroy nature's attempts to bring back the plants. Phillip Island off Norfolk Island is a classic example of how rabbits can create their own desert. As well as rabbits, camels, donkeys, goats and horses disrupt the regeneration of native plants. In addition, there are all the forgotten grazers such as insects. Usually there is a bigger biomass, or weight of animals, below the soil than above it.

Can the land be grazed rather than mined, as has been the pattern for our first 200 years? Scientists are working on this problem and trying to get agreement from governments on the main principles of looking at the health of the country. Is the land improving, static or deteriorating? With the carrying capacity of the land worked out for various conditions in terms of rainfall, it should be possible to rationalise our land use.

This means that owners can no longer regard the land as personal property. None of us can treat the soil as though we own it. We are trustees for the future and if we are not willing to take good care of this priceless asset, legal processes should force this duty upon us.

Causes of soil erosion
Wind

It is amazing how much soil the winds can shift once the covering blanket of plants and plant debris is removed. When our arid and semi-arid country is gripped by drought the bare soil begins to blow. Cities like Melbourne and Sydney are covered by clouds of dust which blanket the sun; particles may reach as far as New Zealand. More common is the shift of soil from paddocks to fence lines or, in the case of our sandy coasts, the movement of the dunes to overwhelm nearby forests.

To keep wind erosion under control we need to look after the plant blanket. This means not only the trees, shrubs and grasses but even the lichens that form a protective crust over many soils in arid Australia. As a side benefit much of our wildlife depends on our keeping this soil cover in good heart. The hard feet of sheep, by chopping up ground litter and the soil surface, can cause the disappearance of trapdoor spiders that rely on this layer both for building and hunting! Trapdoor spiders may not seem important, yet they are part of the complex web of life.

Sheet and rill erosion

Farmers often wonder why stones appear like mushrooms in their paddocks. My father believed the stones actually moved up through the soil. He never realised it was the topsoil washing away that revealed the stones! The tiny hammers of raindrops help splash soil downhill. Water pouring over the surface during heavy showers also carries soil away and the surface layer may be lost. Sometimes the water concentrates into small streams which we know as rills.

Proportion of rural land subject to wind erosion

Source: Prepared from data in Woods, L. E., *Land Degradation In Australia,* AGPS, Canberra, 1983, and Department of Arts, Heritage and Environment.

Gully erosion

When these rills deepen, erosion becomes spectacular. Such gullies are dry most of the year but after heavy rains the soil they are carrying moves into the rivers and finally into the ocean. City dwellers become aware of the problem when their estuaries and beaches become silt laden and swimming is unpleasant.

Tunnel erosion

There is a more subtle kind of wasting when the topsoil forms a hard crust. Should there be a break in this cover the wash goes into the subsoil and gradually dissolves, thus leaving behind tunnels. The eroded soil appears at lower levels, at the bottom of hillsides. When the crust collapses the tunnels are revealed.

The result of all this erosion is that between 20 and 380 tonnes of soil per hectare of affected land can be lost every year. The heavy rainfall regions of Queensland suffer the greatest losses and it has been estimated that four centimetres of topsoil is swept away each year. We may well ask what is blowing in the wind—it is our future!

What should be done?

There are obvious solutions such as conservation farming, where instead of deep ploughing farmers leave a cover of organic material on the soil. Furrows can be made along the contours of the land to stop water from running down the slope taking the soil with it. The keeping of plant cover on the soil is vital.

The land survey already mentioned recommended that attention to soil erosion should go first to those places where there are good soils and adequate rainfall or irrigation. There is an essential need for education in the form of advisory services for all who care for the land. This includes farmers and those who are involved in rivers and estuaries as well as beachfronts. We need to look at the long-term future of our arid lands.

What will it cost?

Although expert opinions vary, the figures range from $675 million to $1 billion in the next 30 years. It is accepted that the work should be carried out on a whole catchment system, so that the problem will be solved over a region rather than by a piecework approach. The basic organisation is in place both at federal and state levels. The experts are available; all we need now is the will!

Mulga and mallee

These two interesting habitats have severe erosion problems. Mulga is found over a vast area of arid Australia and mallee across the southern half of our continent. The clearing of the thickly packed mallee trees soon created problems as the light soils of this low rainfall region were very susceptible to wind erosion.

Mulga country is important for stock grazing. A study in Queensland highlights the problem. Low soil fertility is particularly noticeable in arid areas. The following table compares our soils with those of the rest of the world:

	Nitrogen (% total N)	Phosphorus ppm (ppm)
Australia	0.06	240
Other	0.11	710

As 35 per cent of the mulga country has been found to be susceptible to erosion, the danger is obvious.

Salt

Naturalists can be forgiven for giving a wry smile when reading history. There are tales of the rise and fall of empires, the reasons usually given being invasion by warlike tribes,

Soil salinity in Australia resulting from human activity

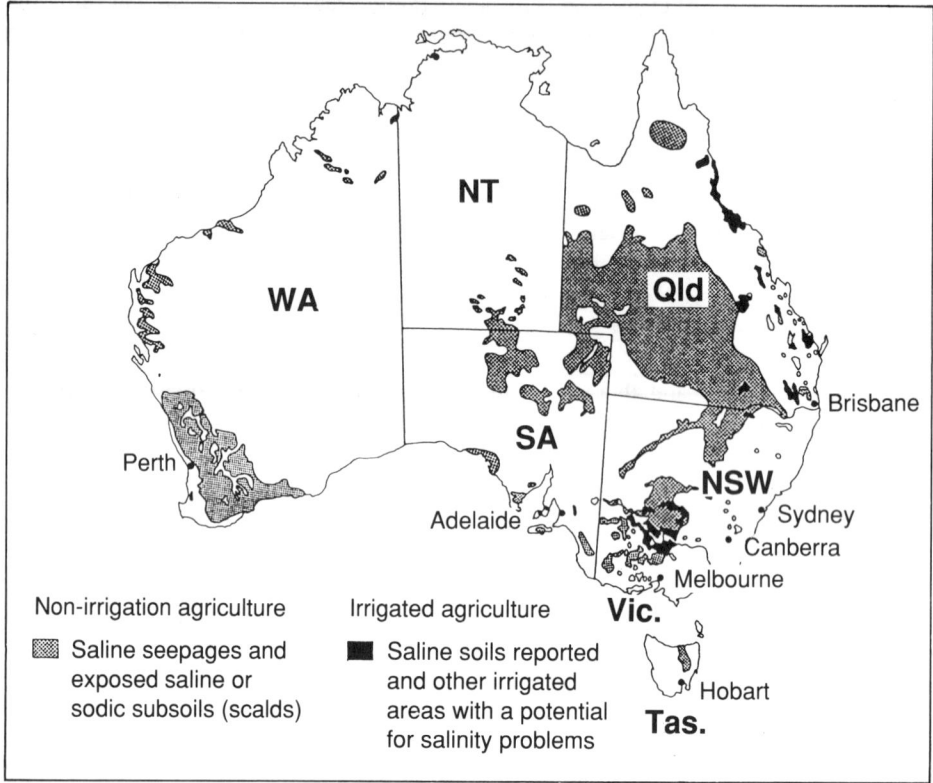

Non-irrigation agriculture
  Saline seepages and exposed saline or sodic subsoils (scalds)

Irrigated agriculture
  Saline soils reported and other irrigated areas with a potential for salinity problems

Source: Peak, A. J., *Salinity Issues, Water 2000*, Consultant Report No. 8 AGPS, Canberra, 1983, and Department of Arts, Heritage and Environment.

Goths, Vandals, Tartars and the like. The more we study the past the more we realise that often the invasions were successful because of other factors. The relentless nibbling of goats has destroyed more fertile countries than war. The strong empires of the Tigris and Euphrates, the Middle East, India and China were built on vast irrigation schemes. Such schemes over hundreds of years caused a rise in salt that destroyed fertility and weakened the civilisations.

The one exception was Egypt, but in the past the lands of the Nile were never irrigated. They were flooded each year and with the water came a fresh load of silt. Any salt was carried down into the ground and from there went into the riverbed. Now that the Aswan Dam has been built, Egypt has to face the problem of salt accumulating in the irrigated fields.

Australia suffers from salinity in agriculture, not only in irrigated fields but also in ordinary dryland farming. In south-western Australia about 4 per cent of cleared land has become salted. In South Australia, Victoria and New South Wales irrigated lands are suffering.

Another concern is the rising salt level in many water supplies used for human drinking water. The World Health Organisation has set 500 milligrams of salt per litre as the highest desirable level with 1500 milligrams per litre as the maximum permissible level for domestic use. A survey of public water supplies in Western Australia showed that 66 per cent are below the 500 milligrams per litre, 95 per cent are below 1000 milligrams per litre and 5 per cent are above this level.

South Australia, which obtains most of its drinking water from the Murray River, has severe problems of high salt levels. This river is our largest and drains more than a million square kilometres, about one-seventh of Australia. About 30 per cent of Adelaide's water supply comes from the river in normal years, but in dry times it may rise to 80 per cent.

How does the salt arrive in our soil?

Much comes from the rain. CSIRO figures show that in coastal regions about 160 kilograms per hectare of salt falls with the rain every year. Further inland the levels drop to about 15 kilograms per hectare at a distance of 150 kilometres from the sea. Salt is also abundant in materials laid down under the sea and such rocks may be even saltier than seawater.

Cutting down the water pumps

Normally, salty groundwaters remain deep in the soil and do not cause any damage to plant roots. The clearing of the trees, nature's water pumps, is the cause of the rise in these groundwaters.

The mallee trees which once covered so much of southern Australia had root systems as deep as 18 metres. Mallee and similar trees literally 'suck up' the water into the leaves where it is transpired into the air; they work therefore as efficient water pumps. Many years ago the value of eucalypts as water pumps was realised overseas. The draining of the Pontine Marsh near Rome, which defied all the engineering skill of the ancient and modern Romans, was achieved by the planting of eucalypts, particularly the river red gum.

Once land has been cleared and planted with short-rooted crops, the salty groundwater begins to rise and low-lying areas become saturated with salt. In the Murray River region the change in the water balance has also forced more saline water into the river system with devastating effects further downstream.

Many schemes are being tried to stem this steady increase in salt in the river. Some are mechanical, like the use of salt ponds; some ecological, like finding plants which can handle salty soils. In irrigation areas new citrus and grapevine are being developed for planting. Another solution is to plant trees on salt-affected land and start the 'pumps' working once more. Two scientists gathered seeds of river red gums growing in salty soils. They found some which could grow in water half the saltiness of seawater.

In the past, growing sufficient individual trees has been difficult; however, today's new techniques of tissue culture have made a significant difference. Thousands of seedlings have been exported to countries such as Israel to help repair salt-damaged soils. Scientists are now searching for eucalyptus, sheoaks and wattles growing in saline areas anywhere in Australia. Not only seeds are collected but also samples of the soil around such plants since these contain the root fungi needed by the plants. It may be possible to find other species perhaps even more effective than the river red gum, and tissue technology will enable them to be used anywhere in the world.

Also being searched for are plants which can survive drought, waterlogging, high water usage and high growth rates. There is no question that one solution to the problems of our timber needs may come from species which grow rapidly. It is often forgotten that our eucalypts are already among the most ecologically valuable of our exports, being cultivated in more than 70 countries around the world. Recent years have shown the economic value of our sheoaks and wattles, and Australia is a storehouse for many nations. Part of our conservation strategy must be to keep all our diversity so that more useful species may be discovered in the plant life gene pool on which we all depend.

Usable agricultural land

Australia's total area of about 780 million hectares has climatic, landform and soil problems that reduce the total arable land to only 77 million hectares. This is one limit which places an automatic restraint on our future expansion. As population rises the possibility of an increasing export of food and fibre will fall. The temptation will then be to extend agriculture on steeper slopes and on climatically risky marginal lands. There is not only local damage to fear; downstream water quality will suffer. The abandonment of farms and unwise clearing will increase both soil erosion and the number of insect pests such as grasshoppers.

Erosion is one problem. There are others, such as salination, acidification, nutrient loss and soil structure changes. The deserts will expand. Areas of desert scattered over the mainland will increase until some coalesce and move with increasing speed. We will have a repeat of what happened in the Mediterranean several thousand years ago.

The Aboriginal hunter-gatherers had ecological restraints; their population adjusted to the wildlife resources of the country. Yet even their use of fire caused changes in plant life as well as erosion.

Land degradation—what is it?

A study of soil conservation needs undertaken between 1975 and 1977 by the states and the Australian government showed that 52 per cent of land in rural use needed some form of treatment for land or plant degradation. The survey showed that over five million square kilometres was used for agricultural and pastoral purposes. Of this area 23 per cent needed treatment with land management practices, while 29 per cent required conservation works of various kinds together with appropriate management practices.

The arid zone

This area, where rainfall is too low or too unreliable to support regular dryland cropping or sown pastures, covers 70 per cent of the mainland. Some 30 per cent is too dry even for pastoral production. The remaining 3.356 million square kilometres of the arid zone provides grazing for about one-quarter of Australia's livestock.

About 2.5 million square kilometres are not used for agricultural or pastoral purposes. The unused land in this arid zone, and some of the unusable land, is of increasing value for tourism, perhaps our most rapidly expanding 'crop', while much has been allocated to Aborigines whose hunter-gatherer style of living is a wise use of the resource.

Degradation has occurred on 55 per cent of the grazing land in this zone and the most obvious damage is a deterioration of the plant life. On about one-quarter of the degraded areas the plant loss is accompanied by some erosion. On another quarter the damage to plants is accompanied by soil erosion and dryland salting.

Improved range management methods are needed on all the degraded land and reduction of stock numbers is necessary for a great part of it. Resumption or amalgamation of some properties with insufficient productive land would help solve some conservation and human problems. Also, perhaps, we need to revise the heroic image of the cattle kings, whose legacy was often the degradation of their empires.

Conservation measures are now needed on one-third of all arid grazing lands.

The non-arid zone

In this area, 30 per cent of Australia, the rainfall normally sustains dryland crops and sown pastures. More than 98 per cent of Australia's population lives in this zone, which also carries about three-quarters of our livestock.

Land which is cropped with varying frequency, and most of it without irrigation, at present is 6 per cent of the area of Australia. Agricultural uses of this land can be classified as grazing, extensive cropping and intensive cropping. More than one-fifth of the zone is forest, parks and reserves, urban areas and unused Crown land.

Probably no more than 10 per cent of the total area is capable of being cropped without an enormous and impracticable expansion of irrigation. In this respect our land resources are very limited and the sooner we realise this the better. Perhaps the sooner the world realises this the better, for envious eyes cast on our wide open spaces should be opened to the ecological realities.

Non-arid grazing land

This amounts to 58 per cent of the non-arid zone. Some 36 per cent of it needs treatment for land degradation. Conservation works and practices are needed on 16 per cent, while land management practices, especially pasture improvement, are needed on 20 per cent.

This land use category includes the catchment areas of the water supplies of many towns and cities. It also includes land formerly sown to crops and pastures but now affected by salinity and offering only limited grazing.

Extensive cropping land

Here is the granary of Australia. It is important also to the livestock industry because of the large number of animals that are grazed on crop residues and pasture. Some 68 per cent of it needs treatment for land degradation. Conservation works and practices are needed on 34 per cent of the cropping land while about the same area can be treated by the adoption of improved land management practice alone.

Intensive cropping land

This is a scarce resource. Only 1 per cent of the non-arid zone, or 5 per cent of all cropland in Australia, is used for intensive cropping and 66 per cent of it needs treatment for land degradation. The scarcity of this land emphasises the need for good management. Conservation works with improved land management are needed on 34 per cent. The rest can be treated by improved land management.

Salinity

In the southern half of the continent there are large areas with increased salinity. Salinity is also a major problem in south-western Australia. One report indicated 1673 square kilometres of saltland, with the increase since 1962 being 529 square kilometres. In some local government areas up to 4 per cent of the cleared land was salted. In the worst affected areas, 85 farms have an average of 95 hectares of saltland.

Some of the irrigation areas of south-eastern Australia have salinity problems. Works to the value of $150 million are being undertaken in the Murray River valley.

Vegetative change in arid and semi-arid areas

Degradation has taken two main courses. The first is the deterioration of the perennial shrub and grass layers. Usually this has been associated with overstocking with the problem made worse by prolonged drought. The loss of plants has resulted in increased erosion with sand drift, wind sheeting and scalding. The extent of the problem may not be as great as was stated some years ago, as a number of years of above average rainfall has brought about a considerable regeneration in many places. Yet not all have recovered and it is obvious that the next series of droughts may result in more destruction of both plants and soil.

Another problem in some parts of arid Australia is the already mentioned rapid encroachment of inedible shrubs. Areas that 30 years ago were a relatively open savannah woodland are now dense shrubland under the tree layer. Grasses have been replaced by shrubs and productivity lowered to almost nothing in many places. Soils are poor, rainfall is low and erratic and at this stage successful techniques for shrub control have not been found. The possibility of using fire is being investigated, but often there is not enough grass to fuel such fires except on days of extreme fire danger when a control burn would be hazardous.

Water

It is a paradox that although Australia is the driest of all permanently inhabited continents (Antarctica is the driest although it has the most fresh water in the world locked in its ice sheets), we have more water available per head of population than any other country. This wealth is the result of government dam-building programmes which save the occasional heavy falls for use in drier periods. At times enthusiasm for dam building has amounted to a kind of hydromania; as one commentator remarked, 'when there is an election in the offing, every politician feels a dam coming on'.

All over the world people are realising that, like soil and air, fresh water is one of the most valuable of all resources. However, because it is cheap we hold its care equally cheaply. Most Australians live on the floodplains of rivers. The result has been massive interference with the natural flow of river systems. The need for water for farming and for urban use has produced many changes, some useful, some wasteful.

It is said that the waters of the Rhine pass through six human bodies before reaching the sea, emphasising that polluted water can be purified and reused. We pour vast quantities from our own activities into the sea, so losing it for human use. Indeed waste materials such as sewage often return to popular beaches making them unfit, if not dangerous, for swimming. This is a waste which cannot go on forever.

The real cost of water

Many economists point out that government pricing policies for water mean that users never pay the real costs of storage and supply. All attempts by private interests to move into the field of water storage and production have failed to show a profit. Because of this lack of real costing our use of water has always been wasteful. It is rare anywhere in the world for an irrigation scheme to be profitable if all costs are taken into account.

A review of the real cost of all water supplies is needed so that correct charges can be made and water conservation will become economically worthwhile. Water is an immense resource. The Sydney Metropolitan Water and Sewerage Board delivers more than 700 million tonnes each year. This is more than the total weight carried by the railways and all the transport systems around Australia!

We, as consumers, need education in water use. A study in Perth showed that toilets are the single largest consumer of household water. (A house with two toilets uses 50 per cent more water than one with only one, possibly due to impulse use!) An average family used 298 tonnes a year, 173 tonnes in the house, 39 per cent in the bathroom. Forty-two per cent was used outside in gardens, and a change to native plants could help in saving this resource. Conversion to toilets with both a half and full flush would also be useful. Most toilets use far more water than they need and a quick remedy is to put a house brick in the cistern.

Water quality

Fresh water expert Professor W. D. Williams states that the major problem is not the amount of water available but the quality of that water. He points out that we need a cohesive national organisation specifically charged to undertake water resource investigations in the broadest sense.

Flood mitigation

The popular appeal of flood control and water storage has often resulted in the natural environment becoming degraded. It has also resulted in valuable wetlands being drained to produce poor quality farmland. Studies in various states reveal that about 60 per cent of Australia's wetlands have been destroyed since European settlement. We urgently need to take a new attitude towards our wetlands and treat them as the valuable resource they are.

Water storage and wildlife

Obvious losses from dam building are the destruction of stands of timber and the flooding of prime agricultural land. The loss of Huon pine was one of the arguments against the Franklin Dam proposal while the flooding of prime farmland was an argument against the Tillega Dam in New South Wales. These examples could be multiplied over Australia. Obviously a balance must be struck between the needs for water storage and hydro-electric power and the other uses for the same land and water.

Some damage can be remedied. Fish ladders may provide for migratory fish on breeding journeys. Without these some of our native fish may disappear. The deep waters of dams are unsuited for most wildlife, and some governments have created shallow ponds at high-water level that become wildlife havens when filled.

The giant Ord River Dam in Western Australia does have a large area of shallows that is a haven for waterbirds and fish. This attracts tourists and it is tourism that has produced the largest economic return from this project, as all the 'dream' crops proposed for the area have failed, though there have been some recent minor successes.

Very cold water being released from storage to flow downstream can also damage wildlife and this problem must be recognised and overcome.

River channel management

Many of the mistakes of the past are now being avoided. The problem of pollutants being poured into streams has been solved in part by new laws. Legislation is also controlling land use in floodplain areas. Wetlands are seen to have value both in providing storage ponds during floods and as wildlife habitats. In North America sporting shooters have spent millions of dollars to create artificial wetlands so that duck populations remain high! There is another benefit, even for those who do not enjoy hunting: a lake for ducks is also an attraction for other wildlife. The hunting season is short so for most of the year such wetlands can be enjoyed for more passive recreation.

Effluent released into streams and the sea is a problem in many parts of Australia. We must hope that in the future when the value of water is realised the reuse of waste water will reduce this problem. At present there are few unpolluted streams in Australia. Most are in national parks and similar reserves.

Local authorities tend to turn attractive rivers and streams into concrete channels. River red gum regeneration stimulated by irregular flooding has been affected. The Murray cod needs the summer floods of shallow warm water to trigger egg-laying in submerged logs and other sites. Without these warm shallows the fish does not breed in the numbers needed to keep up the stocks of this important food.

The fight for the Franklin River in south-west Tasmania was the most dramatic of Australia's conservation struggles. Governments must look at all sides of the water storage problem before making a decision on the wise use of this scarce resource.

Salination

The salinity of some of our most valuable soils has been discussed. In the Murray River area irrigation has resulted in a build up of saline water tables and drainage from such areas has increased salinities in the lower Murray. Although some of this salinity is being removed from the catchment into evaporating basins, the salt levels towards the river mouth are still so high that in Adelaide, which depends on this river for its fresh water, the water may be so salty that it tastes unpleasant, and well as being a health hazard.

In south-western Australia and the mallee country of the south-east the clearing of the forests removed the deep-rooted trees which acted as water pumps and kept the saline water table well below the surface. When these natural pumps were removed the salt water rose to the surface and evaporated, leaving behind a coating of salt.

Replacing tree cover is one solution. Obviously this can be done only in seriously affected areas or on the upper slopes of catchments, as farmers must keep the bulk of their cleared land for crops or grazing. Each farmer must decide how best to balance the steady encroachment of the salt against the economic return from the land. In some salt prone areas of Western Australia the government has prohibited clearing to prevent any further increase in salinity.

Other poisons

Pesticides, including herbicides, are polluting streams sometimes with heavy losses in aquatic wildlife. As yet little is known about the effect of increased fertilisers in our streams from farm scour. Certainly this extra nutrient produces excessive plant growth in the water and farms using fertilised waters for stock find the excess plant growth can be poisonous. To take only one example, Colin Creighton of Landwater Management, ACT, makes the following comments:

Managing agriculture to maximise profits and minimise red spot

Recent articles in various journals including the National Parks Association *Journal* seem to be treating the fish disease known as red spot as of unknown cause. Red spot occurs worldwide in a number of estuaries and embayments which have heavily industrialised catchments. However, to put the Clarence [River] in this category is a sad indictment of our land management practices, particularly when it is noted that agriculture is virtually the only industry in the Clarence catchment. Rather than further research, the Clarence and all other barrier estuaries of the eastern seaboard require management of uses such as agriculture. Following is a list of simple management tasks that would both alleviate the red spot problem and increase overall the productivity of our estuaries:

- reduce enrichment loading by the implementation of guidelines for superphosphate applications (grazing lands);
- reduce enrichment loading by the implementation of guidelines for aqueous ammonia applications (cane lands);
- reduce enrichment loading by the land disposal of treated sewage (examples of cost-effective methods include irrigated pastures, plantation forestry and irrigated-fertilised sugar cane as is common practice in Hawaii);
- ban organochlorines and organophosphates for all uses (the Commonwealth should come to the party on this one and ban the import of these persistent poisons);

- control soil erosion, all land uses;
- manage flood mitigation barrages to recreate habitat and control the aquatic weeds salvinia and water hyacinth with brackish water rather than poisons;
- rehabilitate wetlands, particularly the fresh to brackish systems which provide the buffering capacity of our estuaries.

All of these are fairly simple management tasks. All, in their own right, are cost-effective for the agrarian sector without even considering the spin-offs to our estuaries.

Pesticide concentrations in organisms from an American river estuary

Source	Pesticide concentration (ppm)
Water	0.000 05
Plankton	0.04
Plankton-feeding fish	0.02
Predatory fish	2
Gull (a scavenger)	6
Cormorant (feeds on large fish)	26

Source: Biological Science: The Web of Life.

Some scientists claim that similar nutrients may be causing harm to the Great Barrier Reef through runoff from cane farms. The nutrient may increase the water plants, therefore assisting in the increase of the crown-of-thorns seastar whose larval stage feeds on such floating plant life.

Urban areas

About 85 per cent of Australians live in cities or towns and urban pollution is therefore a growing concern. New controls on water pollution have been established in all states and gradually the results are becoming apparent. One example is the Parramatta River which runs into Sydney Harbour. Only a few years ago it was gradually 'dying' as the larger fish disappeared; today, as a result of a conservation campaign, these species are returning.

Sewage pollution is a continuing problem in many parts of Australia, particularly on Sydney beaches where onshore winds often produce disgusting conditions. New ocean outfalls which will discharge the sewage further out to sea should solve this problem, though this treasurehouse of freshwater is still being wasted by ocean disposal.

Bank erosion

A constant problem is the erosion of valuable soil from riverbanks. Often this is due to unwise clearing which removes the soil-binding roots of trees. In the early days of settlement in New South Wales trees such as river oaks and mangroves were protected because of their value as soil binders.

In many streams around Australia the clearing of snags and other obstacles to give rapid runoff and prevent flooding results in the deepening of riverbeds, while the faster flow increases erosion of the banks and downstream silting. In catchment areas this can be a serious problem. In the Ord River catchment, for example, the government removed all stock from the area so that the plant life would be left undisturbed to hold the soil firmly and not allow it to wash into the dam.

The natural pattern is for a river plain to flood in wet seasons and leave the silt on the flats. Improved land use would not interfere with this age-old pattern and would avoid many of our present problems.

The future

Conservationists complain that promises of multiple water use tend to be forgotten when a drought brings demands from farmers that their needs are paramount when water alloca-tions are decided. A long struggle took place to save the Macquarie Marshes in central New South Wales, one of the greatest of our wetlands. Recently the government agreed that the marshes must be given enough water to ensure that this wildlife haven is kept safe.

Consideration must be given to multiple uses when any change in a water resource is being planned, so that full value can be gained from this most valuable of resources.

Self-help

I saw a marvellous example of self-help when I opened an international seminar on wetlands to coincide with the International Wetland Year.

A year before, the Newcastle City Council had planned to use a local swamp for a rubbish dump. This has been a normal pattern around Australia and many playing fields have been created in this way. Many of the residents of the Hunter Valley saw swamps as mosquito ridden and useless. Gradually, conservation groups have been changing public attitudes by pointing out that such freshwater and marine marshes are more valuable (even in monetary terms, five times more according to American figures) as they are than if they were used as agricultural or urban land.

Fourteen residents decided to do battle with the council. They won that fight and the swamp was saved. The group knew that no conservation battle is ever entirely won until the victory is enshrined in legislation, or some other more permanent protection, so they formed the Hunter Wetlands Trust. A year later the 14 original members had grown to two thousand.

They raised $1.2 million to create a vast wetland sanctuary; they bought an old clubhouse and turned it into an information centre and they created a 12 kilometre long canoe trail through a series of wetlands which they now manage. The council which had originally planned to fill in the swamp donated $100 000 towards the project.

The crowning triumph was that a football field, which had been created by filling in a swamp decades before, was dug out to let the waters flow back. A classic case of turning

A wetland near Nowra, NSW—what would the world be without areas such as this?

swords into ploughshares! Perhaps we have learned something from the great American conservationist Henry David Thoreau who said: 'A town is saved no more by righteous men in it than by the woods and swamps that surround it . . . '

Air

Most lifeforms need air to survive. This remarkable material is a mixture of gases—78 per cent nitrogen and 21 per cent oxygen. One per cent is a mixture of inert gases of which argon is the most important while neon and helium are two others familiar to most of us. A small

Mountain ash. These are the world's tallest hardwood trees with heights of up to 98 metres. In most countries such trees would be valuable tourist assets but in Tasmania they are almost unknown, even though these giants in the Styx Valley are close to Hobart. (See also p. 112)

Tasmanian Tigers, a painting from John Gould's Mammals of Australia. *John Gould was the British naturalist who, through his books, made Australian wildlife known to Europeans more than 100 years ago. The Tasmanian tiger was exterminated by being hunted as a pest since it killed sheep. (See also pp. 59–65)*

Author and a hair seal on the beach at Seal Bay, Kangaroo Island. This is one of the few places in the world where a tourist can drive to a beach, walk along a path over the dunes and then join breeding hair seals on this beach. All marine mammals are protected in Australia. (See also pp. 59–75)

The brown haze which lies over Sydney, viewed from North Head at the entrance to the harbour. (See also pp. 32–6)

but vital part (only one-tenth of 1 per cent or 330 parts per million) is carbon dioxide. Water vapour may vary from 0.009 per cent to 0.9 per cent; it is more humid near the coasts and drier in the deserts.

Then there are the pollutants, some of which are natural phenomena. Thunderstorms create oxides of nitrogen through the energy of the lightning flash, chemically combining the oxygen and nitrogen in the air. Many plants emit volatile hydrocarbons—it has been estimated that each year the world's plants give out nearly one thousand million tonnes. The smell of the eucalypt forests is due to this and so is the blue haze which creates our 'blue mountains'. These substances help form natural 'acid rain' and may in the past have contributed to the creation of some of the earth's oil resources.

When the first Europeans settled in Australia the only air pollution would have been from smoke from Aboriginal fires and dust brought from the inland by strong winds. Dr Greg Ayers of the CSIRO estimates that natural sources (bushfires and soil) contribute 75 000 tonnes of sulphur into the Northern Territory Top End air annually.

The next 200 years has seen a steady rise in those types of pollution and an increase in many new forms, a trend found around the world as new demands and new technologies put pressures on environmental quality. Modern technology has increased urban pollution of this kind. Other problems include disturbance of the ozone layer, aerosols and small particles floating free.

Unfortunately, during the past 200 years people have gradually become accustomed to bad conditions, although there has been a significant change in the last 20 years. Increasingly, the public has demanded a return to clean rivers, lakes, seashores and air.

In Australia a number of enquiries at federal, state and local government level have revealed widespread pollution of many kinds. Non-government organisations have also played an important part in uncovering abuses of the pollution laws, leading not only to prosecutions but also to a strengthening of legal controls.

The worst pollution concentrations have been found in cities and near mining and manufacturing regions, and there are specialised types of pollution in food processing areas. In places of intensive farming, heavy use of pesticides has caused problems, the effects of which are widespread due to movement of the poisons by both air and water. For example, some coral reefs are being polluted by runoff from cane farms. Cotton farming has produced heavy levels of pesticides which have aroused alarm, about not only the danger to wildlife but also the danger to people. Many such poisons may cause cancers and this is a new cause for concern. Increasing erosion creates more wind-moved dust in the air, while severe bushfires produce smoke problems at particular times.

These problems are found both in Australia and New Zealand. In the latter country the Clean Air Act was passed as long ago as 1972 and the Water Pollution Control Council is responsible for controlling the quality of natural waters. Tourism is important in both countries and is likely to become a very large income earner so water quality and clean air are also important. Sport fishing in rivers, lakes and seas is of prime interest to many visitors and the cleanliness and open space of both countries is another major attraction.

Modern science and its associated technology has been responsible for introducing new forms of pollution. Synthetic chemicals have been used on a large scale and many have escaped into the environment. The most publicised of these is DDT. Its dangers were highlighted in *Silent Spring* by Rachel Carson, who created world awareness of the dangers posed by non-biodegradable pesticides. After killing the pest, the poison continues along food chains, destroying in its progress harmless or useful forms of wildlife. The title *Silent Spring* reflects the fact that birds were among the more obvious wildlife victims.

The combined effect of increasing consumer demands due to affluence and a rising world population has brought new pressure on the environment. The need for more food to be produced in the same area of agricultural land has brought about the development of new chemicals to control plant and animal pests.

Although in Australia there has been a gradual phasing out of the pesticides which do not degrade quickly, there is still need for improvement. New fears have arisen that some of these poisons are not only killing pests but are affecting the health of people and in particular causing cancers.

The irony of pesticides is that the more we use, the more pests appear on the agricultural scene. There are a number of reasons for this but the major one is that pesticides kill the innocent as well as the guilty, the predators of the pest as well as the pest itself. Sometimes removing one pest gives another minor insect the chance to fill the gap. In the huge numbers of any pest species are varieties which are immune to the poison. Today nearly 500 kinds of insect pests are immune to particular pesticides. Some are immune to all.

So the new drive is for integrated pest management and also research on new poisons. If we think laterally, might it not be that since the more pesticides we use the more pests we have, then the fewer pesticides we use the fewer pests we may have!

An American ecologist, Gary Barrett, is experimenting with an old idea; the setting aside of a two-metre-wide corridor of wild land around every two-hectare paddock. In this strip of native vegetation, enemies of the pests flourish year round, and survive when there are no crops to attack. Such a strip is also a barrier to those insects that at present can move steadily over the huge monocultures of modern farms.

We are beginning to learn of the substances plants put into the air. Perhaps these chemical messengers may confuse pest species, adding to the barrier effect. Such native bush helps solve other problems such as wind erosion and provides shelter belts. When I pushed for this idea of wildlife corridors some years ago, I called them 'koala corridors'. If farmers and urban conservationists both planted trees for koalas, the same plants would serve other wildlife as varied as sugar gliders, black cockatoos and wasps which feed on the nectar of tree blossoms to sustain them in their battle with chafer grubs in the paddock. The loss of crop area to sustain these corridors will be repaid to the farmer not only in more economic production but also with the pleasure of the wildlife the strips will sustain.

There has been increased government interest in integrated pest management where a broad spectrum of controls are used, including new farming methods and biological controls similar to the attack on prickly pear with the cactoblastis caterpillar and the use of the myxomatosis virus on rabbits. If this approach is adopted, pesticides will be used as a last resort or in conjunction with other controls. Such management is also useful in cutting farm costs.

New understanding and manipulation of genetics is leading to the breeding of higher yielding and more pest-resistant crop plants and farm animals so the future looks brighter than it did 20 years ago.

Sydney smog in the early 1970s. Since then, the introduction of emission controls has resulted in a dramatic improvement. (Sydney Morning Herald)

Motor cars

Air pollution caused by motor cars has been added to that from industrial smoke stacks, smoke from backyard burning, commercial incinerators, wood stoves for warming houses and other urban discharges from a variety of sources. Of all these factors the motor car has been the biggest problem.

Pollutants from cars include:

Lead additives These create more efficient fuels but the lead passes into the air where it can be inhaled. Children are the greatest sufferers as lead can damage brain cells, reducing intelligence and concentration. This effect is compounded in some areas where lead fallout from smelters is added. New South Wales has been the leader in introducing lead-free petrol and in a few more years all Australian states will follow suit. This should remove lead poisoning as a major pollution problem.

Carbon monoxide A colourless and odourless exhaust gas, it is highly toxic and at busy intersections in cities can reach dangerous levels.

Hydrocarbons Some fuel is not burned in car engine cylinders and emerges into the air as an irritant, although it is not dangerous to humans.

Oxides of nitrogen These gases irritate people and, combined with rainwater, produce acids which are corrosive and can destroy buildings and other surfaces.

Photochemical smog This word has become well known since deaths have been caused by air pollution in cities such as London and Los Angeles. The main problem is chemical reactions produced by sunlight interacting with pollutants such as hydrocarbons and oxides of nitrogen. The end products cause respiratory problems, headaches, sore eyes and throats, nausea, reduced visibility, damage to plant life and also the deterioration of material such as rubber and clothing.

Smog has been reduced by stringent emission controls on new vehicles. The smog problem is found in most large Australian cities, particularly if they lie in hollows or are backed by high mountains which trap pollutants. One example of the danger of smog was shown dramatically in one incident in Sydney in 1976. Thirteen children taking part in high school sports suffered chest pains and breathing difficulties and had to be taken to hospital. The more gradual ill-effects and lessening of life expectation in older people are not so easily measured.

Smog is usually produced in the warmer months from October to April. Clear skies, light winds and an afternoon sea breeze, conditions usually regarded as desirable, are also ideal for smog. One of the major products is ozone and this offers a simple measure of smog levels. Smog also produces a grey haze.

Emission controls have produced dramatic results in Sydney. Since 1976 when they were introduced there has been an 80 per cent drop in ozone. The high smog days have also fallen from 20 a year in the mid-1970s to about six in 1986. There has been little ozone damage to plants.

Dust levels have also fallen by 60 per cent from about six grams a square metre a month to between two and three grams. Acid levels in rain have been halved. Similar results are being obtained in other cities.

Brown haze This has varied sources including sea salt, motor vehicle emissions, backyard burning, smoke stacks, incinerators and some soil dust. Backyard burning has now been brought under control in Sydney.

Monitoring pollution

At Cape Grim in western Tasmania an Australian Baseline Monitoring Station carries out a continuous testing of the air coming in on the Roaring Forties. The air here is as pure as is possible in Australia. Any world increase in a pollutant will be detected, and the air purity at Cape Grim can be used to check the air in our cities. As well as the stationary monitors in all major cities, some mobile stations are used for special cases.

It is a truism that there is only one earth. Nowhere is this better shown than in the question of global air pollutants.

The ozone layer

Most of us admire the blue skies above us and enjoy the tanning effect of the rays pouring on to ocean beaches. We forget that high above is a very thin layer of ozone which protects us from getting too much of a good thing. An excess of those same rays can kill us and also affect plants on land and in the sea. It is therefore important that this layer be kept safe.

Some years ago there was great alarm when it was found that this shield was being eroded by the propellant used in aerosols, those convenient household sprays. These are the chlorofluorocarbons. Since then we have discovered that many other pollutants are being produced which could affect the ozone layer. Polystyrene foam packs when burnt are one problem and so are the host of other materials used in foam ceilings, refrigerated trucks and in the cleaning of electrical circuit boards. It appears that this vital ozone layer is being eroded at the rate of 0.05 per cent annually.

Fortunately, international action is being taken. In 1985, 28 nations including the USSR signed the Vienna Convention for the Protection of the Ozone Layer. In 1987, 40 countries signed the Montreal Protocol to reduce chloro-fluorocarbons, a clear signal of how advanced nations realise the danger. Perhaps those who complain of the increasing

Chloro-fluorocarbon levels

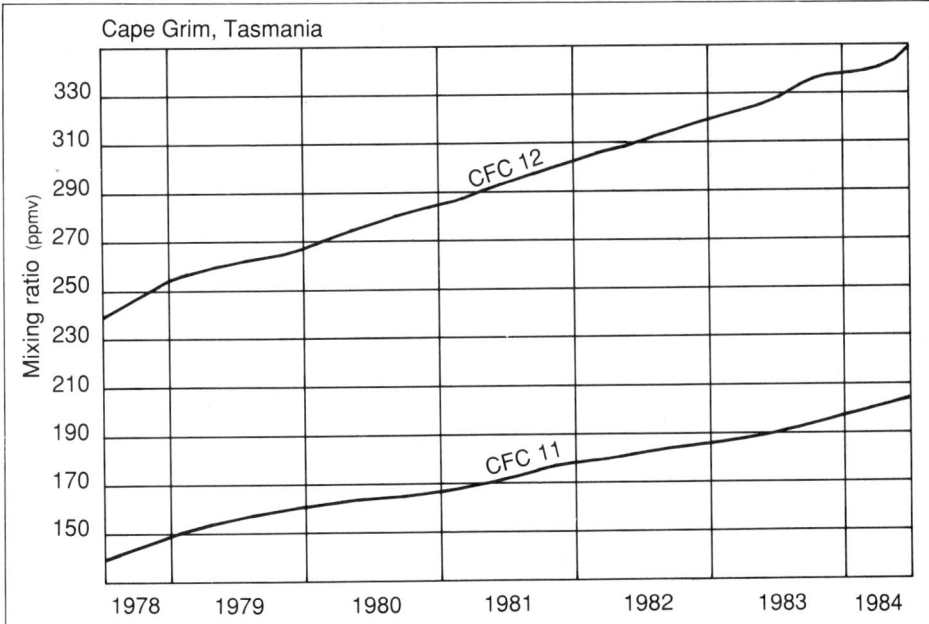

Monthly measurements of concentrations in the atmosphere since 1978 show a slow but steady rise. Because CFCs remain in the atmosphere for as long as 100 years, it would be a long time before decreases in CFC production caused atmospheric concentrations to fall.

Source: UNEP.

Chloro-fluorocarbon production

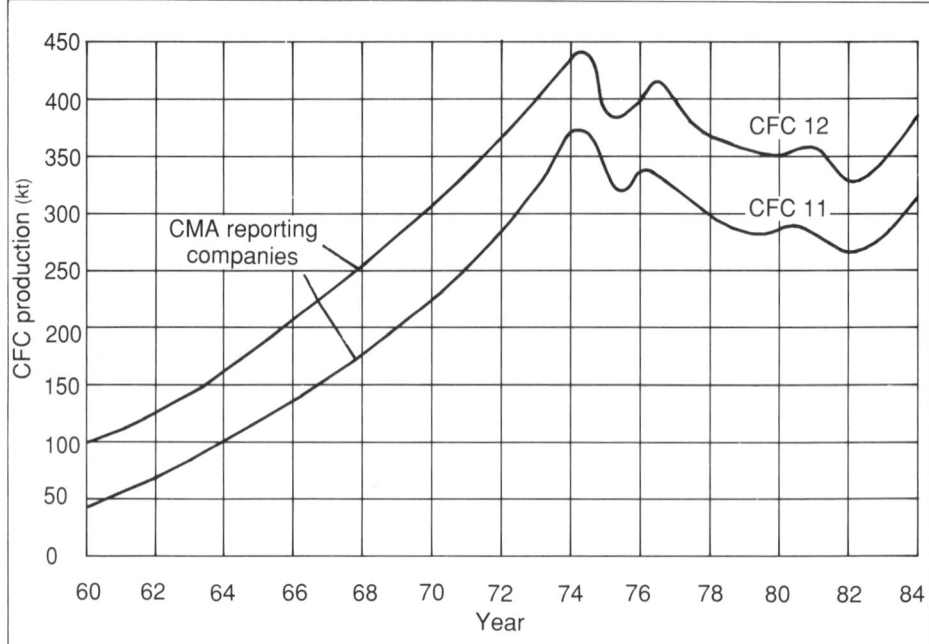

Production of CFCs 11 and 12 has increased substantially since 1960. After a slow-down in the mid-1970s, production appears to be increasing again. Graphs are compiled from statistics supplied by the Chemical Manufacturers' Association (CMA).

Source: UNEP.

interference of governments in the private affairs of citizens should take note of such conventions and be grateful! Often the interference is to stop profiteers from robbing us, poisoning us or, most dangerous of all, creating problems in this environment on which we all depend, even the profiteers. A wise government must take steps to protect people from their own follies!

Carbon dioxide

This gas, essential to all life on earth, has caused alarm in recent years. The balance of heat on the earth's surface is due to the high-level energy from the sun passing with little loss through the atmosphere and striking the earth where it produces the kind of climatic conditions with which life can cope. A great deal of this heat is radiated back into space but, being at a lower energy level, much of it is trapped by the atmosphere particularly when there is a heavy cloud layer.

This is known as the 'greenhouse effect' as it was observed many years ago that a building with glass walls retained the heat from the sun. The glass allowed direct energy from the sun to pass through almost unhindered but held back the heat radiated from inside the greenhouse.

Carbon dioxide acts like the glass in creating a partial heat blanket to keep the heat energy trapped in our soil and air. Any increase in carbon dioxide levels could increase the amount of heat on the earth's surface and atmosphere. Studies show that carbon dioxide levels have been increasing steadily in this century.

A certain level of this gas is essential, as the combining of carbon dioxide with water by the 'life stuff' of leaf green or chlorophyll using the energy of sunlight, is the essential base of the pyramid of life on earth. About two-thirds of the increase in the gas comes from the production of electricity from fossil fuels such as coal, oil or natural gas. These fossil fuels represent carbon dioxide locked away in the earth over hundreds of millions of years.

Should carbon dioxide levels continue to increase at the present rate it is possible that there could be a warming of the earth. This would lead to the melting of some of the polar ice caps and a rise in sea levels. This in turn would not only flood the larger cities of the world, but lead to a fatal loss of the most fertile of our farmlands.

Can we do anything about the greenhouse effect? It will not be easy. The best method is to reduce the use of fossil fuels (coal, oil and gas) for energy. We need to use other sources of power such as tidal, geothermal and solar energy. Great strides are being made in better design of homes and buildings to cut the need for artificial warming and cooling. Research into more efficient solar cells is continuing.

A great deal of the carbon dioxide now in the air was once locked in the trunks of forest trees. The cutting of forests around the world for woodchips has increased the amount of carbon dioxide as the paper made is usually burned as waste. More recycling of waste paper would help solve this problem.

Finally, when people realise that increasing our consumer needs steadily year by year does not increase happiness, the closer we will come to a steady economy. This is nature's way and will not mean a lower standard of living but a higher one.

World temperature

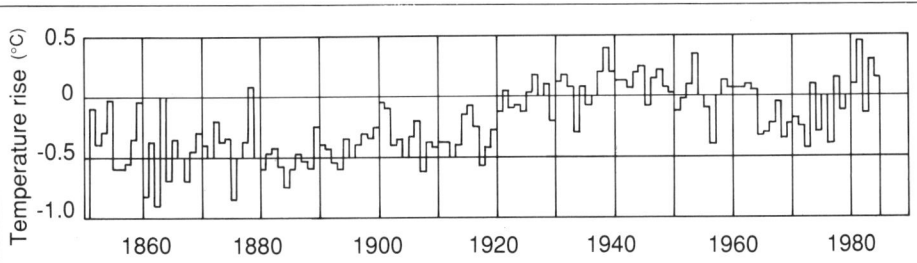

Average annual temperatures of landmasses in the northern hemisphere since 1850. Three phases can be distinguished: a slow rise to 1940; a sharp fall to about 1970; and a rapid rise to the present day. The overall rise—of some 0.5°C—is thought to be at least partially due to greenhouse warming.

Source: UNEP.

Carbon dioxide levels

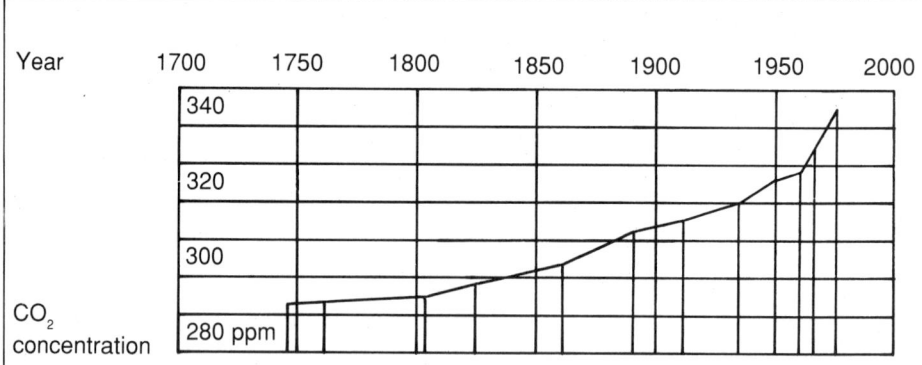

The analysis of air trapped in ice preserved since the eighteenth century shows that carbon dioxide concentrations began to rise early in the last century—and have continued to do so ever since.

Source: UNEP.

Acid rain

The oxides of sulphur and nitrogen can combine with water to make weak acids; weak in chemical terms, yet strong enough to gradually dissolve the stones of old buildings in cities and capable over time of killing plant life. In some European countries as many as one tree in three is dying from the effects of acid rain.

Sulphur dioxide is often produced by the burning of fossil fuels although, fortunately, Australian coal, oil and natural gas have a low sulphur content so we do not suffer as much as other countries. Some industrial processes, particularly metal extracting, produce large quantities of gases which can become acids when dissolved by rain.

The blasted slopes near Queenstown in Tasmania are a classic example of the damage to plant life caused by acid rain. People helped too, by cutting down the larger trees to fuel the mines, and now this landscape looks rather like the surface of the moon and is a chilling example of the earth's future if we cannot cope with this new form of pollution.

Recent research shows that oxides of nitrogen also play a part in creating acid rain in European cities, and for these car emissions are blamed. Should we need to increase our imports of fossil fuels, acid rain might become a problem in Australia.

Minor pollutants

Noise pollution   Sydney has taken a lead in controlling noise pollution through the Noise Control Act. Its operation is based on five controls—scheduling and licensing of premises, prohibition of the sale of noisy vehicles, and the issue of noise control notices, noise abatement orders and noise abatement directions.

Electromagnetic radiation   This is a more recent danger from technology as varied as

Acid rain from metal smelters and overcutting of timber for mines has produced this wasteland at Queenstown, Tasmania.

microwave ovens, radar and high tension transmission lines. Governments are introducing safeguards in these areas but recent studies indicate that there is a link between exposure to these fields and childhood cancer. A conservation group in Victoria has recently been working to stop the building of high voltage lines near schools.

So, air pollution is not something of interest only to scientists. Extinction may even come if a rising of sea levels, through the greenhouse effect, produces such stresses in a heavily populated earth that the result is nuclear war.

Legal controls

In food manufacturing, mining and industrial production there are increasing legal controls on the pollution emitted into the air, rivers, lakes or seas. One example in Australia is electric power generation which mainly uses pulverised coal for fuel. About 33 million tonnes is burned each year, producing three million tonnes of waste material (fly-ash). Most of this dust is trapped by electrostatic precipitators and the amount allowed to escape into the air is controlled by law. Research by the CSIRO has now produced a rise in efficiency and the previous capture of 86 per cent of fly-ash has risen to 99.5 per cent. The trapped waste material is also being used in cement and brick making.

Controls have been applied to oil pollution with both local and international obligations being governed by strict laws. There are still accidental disasters when tankers run aground and there is pressure to force either the use of double hulls for tankers so a grounding does not result in oil spillage. The banning of tanker movements through shallow seas, such as those of the Great Barrier Reef region, would be a wise step.

How many Australians?

Wider still and wider
Shall her bounds be set. . .

During this century, along with the rise of technology arose a comfortable belief that the earth was an inexhaustible source for all our needs. Should we run short here, we could colonise other worlds. Such a belief may be useful for science fiction writers but has no validity for planners of our future.

The unfortunate lemmings were driven by overpopulation to make dramatic cross-country movements which sometimes led them to the sea's edge rather than a riverbank. Their plunge into the water was not mass suicide; they believed that the other bank would be a land filled with plenty. They did not know the other bank was thousands of kilometres away. Australia's long-haired rats sometimes show the same behaviour pattern. If people behave like lemmings or long-haired rats, we will suffer the same fate. We must learn to live within our means; there is no other 'bank' close to us in the oceans of space!

World population

Today this stands at about 5 billion and indications are it will grow to 6.12 billion by the year 2000 and peak at 10 billion by 2080. Superficially, there appear to be surpluses for all human needs, including fossil fuels and food. What is not realised is that if the rest of the world used resources at the rate of Western countries, all those surpluses would vanish like snow in summer.

An Australian uses 50 to 60 times more resources than a person living in the developing world. We claim that many of the world's non-renewable resources will last 50 to 100 years but when they are divided among all humanity, the end of the road becomes dangerously near. There is no safety margin for renewable resources such as food.

What of Australia?

Our numbers grew from 7.6 million in 1947 to 15 million in 1982. About two-thirds of this rise was natural growth and one-third from immigration. Our growth declined from two per cent to a little more than one per cent in the 1980s. Population experts calculate that it may rise slightly to 1.7 per cent in the second half of the 1980s, then fall to about 1.2 per cent by the year 2000. Various projections of our total population place it at 18 to 20 million by the end of the century, with Western Australia, Queensland and the territories showing the biggest rises. At present the five major cities (covering less than 1 per cent of the land) hold 45 per cent of the people. There has been a move away from cities by those who can afford it and a move to cities by the poor and helpless.

How many Australians should there be? This is hotly argued. Patriotic songs see no limits but a study by R. M. Gifford, J. D. Kalma, R. Aston and R. J. Millington came to a different conclusion by examining food and water restraints on our future. If we continue exporting food at our present rates, the maximum, sustainable population would be 22 million. If we eat all the food we produce then the supportable population would be 60 million. These figures have been challenged on a number of grounds, one being that there would not be enough fresh water to sustain such numbers.

For those who feel it is our moral duty to absorb the hungry of the world, it is worth noting that one year's increase from India or Africa would take our population beyond safe limits. Even if the figures are approximate they help to dispel the myths developed by those who, confused by the size of our continent, think we can support the same population as an area such as Europe or North America.

Ecological realities

Our present high standards of living can be sustained only if we remember the restraints mentioned earlier in this chapter. Not only is our population rising steadily, so too are our demands on resources. In the long run such an increase cannot continue. Our needs may tempt us to try and increase our food production beyond the limits our soils and water supplies can sustain. There is always a temptation when demand is high to press on towards such limits. The famous comments by former Victorian Premier Henry Bolte, 'What's a little pollution compared to a million dollar industry?' and the mayor in north-west Australia who said to a ratepayer complaining about dust storms 'who cares about the dust so long as it doesn't clog the cash registers!' show that those who care only for short-term gains are the real danger to our future.

Our waters, soils, air, and plant and animal life have all suffered in the past but often through ignorance. There is no longer an excuse to repeat the mistakes of the past. The NCSA is intended to give sound guidelines for a prosperous future. It must be remembered that we can increase our standard of living if we think in terms of increasing our leisure, our educational skills, our service industries, and all those other activities which make life more fulfilling but do not depend on an increase in the number of things we possess.

On a much bigger Australia

Prime Minister Robert Hawke was quoted in *The Australian Magazine*, 1–2 December 1989: 'It's my personal belief that a larger population, rather than a smaller population, is going to be for the benefit of Australia long-term...The bigger the population, the better in economic terms because the bigger the domestic market. There are economies of scale, your competitive unit cost position improves.'

Almost every politician since Federation has preached the doctrine that bigger is better! By the same logic China should be best country in the world, followed by the USSR and then the US. Conversely Sweden and Switzerland with less than 10 million each should be inferior. In actual fact both of these small countries are world leaders in terms of technology and quality of life. And what of the ancient Greeks? With city populations in the tens of thousands (hundreds of thousands at the most) they achieved standards of art and politics the envy of the modern world.

3. How to use our land

Ensure that productive agricultural and forestry systems are used on a sustainable basis.
Ensure that land which is suitable for many sustainable uses is used in a manner
which retains, as far as possible, the greatest number of options for future use. (NCSA.)

Without natural resources the most noble of aspirations fade and die. Bernard Shaw made
this point with typical logic and wit in his play *Major Barbara*. One of his characters pointed
out that before you can begin filling human minds with great thoughts and great dreams, you
must first fill their bellies.

We *can* have our cake and eat it. That is what conservation is all about—the management
of our resources so that we can use them forever.

Reserve good cropland for crops

In view of the scarcity of high-quality arable land and the rising demand for food and
other agricultural products, land that is most suitable for crops should be reserved for
agriculture. This will reduce the pressure on ecologically fragile marginal lands which
tend to degrade rapidly if exploited beyond their productive capacities. However, this
requirement may conflict with urban, industrial, energy or transport policy. There are
many examples of prime farmland drowned by dams or lost to airports, roads, factories
or housing. Without careful planning and zoning, human settlements sited in farming
areas are bound to encroach on farmland as they expand. Such conflict should be
anticipated and where possible avoided. Since it is not possible to re-site high-quality
cropland but it is possible to be flexible about the siting of buildings, roads and other
structures, agriculture as a general rule should have precedence.

(World Conservation Strategy)

Population and people pressure

Two aspects of modern life press hard on the earth. First there is the rise in consumer
demands produced by technology and affluence. Many estimates have been made on
resource use in the Western world compared to that of developing countries. The diagram
from the WCS shows this clearly.

Such consumer pressure not only exhausts a nation's own resources, it also leads to the
rape of developing countries to obtain raw materials, an exploitation which began thousands of years ago but rose to new heights in the last 100 years. Worse, it also distorts the
development of those countries, often forcing the people to abandon growing their own food
in favour of goods which can be sold abroad to bring in the cash needed for other
developments, particularly the buying of the manufactured goods of developed countries.
'New lamps for old' is the siren cry.

Disproportionate consumption of resources by the affluent

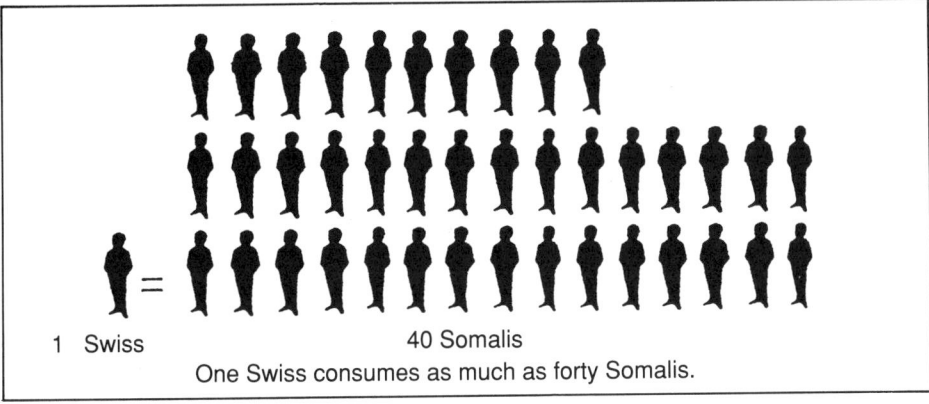

1 Swiss   40 Somalis
One Swiss consumes as much as forty Somalis.

Source: World Conservation Strategy.

Two examples will highlight the irony of this. Australia and Japan are increasingly taking greater care of their own forests yet companies located in both countries are encouraging destruction of the forests in South-East Asia to supply their needs! Even more infamous is the 'hamburger connection'. American companies clearfell the vast rainforests of the Amazon into woodchips for paper. Once cleared, the land is used to grow lean meat for overseas markets.

This demand on the resources of the developing world keeps escalating, as individual consumer wants, fuelled by advertising, rise. It is time for us, both in terms of our own lives and also in governmental demands, to consider wants and needs. We want many things we do not need.

The dream of idealists 200 years ago was that every person should have a chicken in the pot every day of their lives. Today we have chickens in abundance but the wants have risen to two cars in every garage, a radio and television set in every room and a host of other luxuries. So what? This is a common retort of the affluent Western countries in which the extravagances of millionaires are admired and envied and in which the media continually puts forward the illusion that, with a little luck, anyone may join the select few at the consumer 'trough'.

Yet if every Ethiopian and Somali, and the people of other developing nations, lived at our present standards, all the so-called abundance of food, metals and the like would vanish in a few decades. There are not enough resources in the world to support even the present world population at the highest of present living standards.

If land degradation continues at the present rate, close to one-third of the world's arable areas will be destroyed in the next twenty years. Similarly at the present rate of logging of tropical rainforests this resource will be halved in the next twenty years and halved again by 2020.

Television watchers, with tears in their eyes, view the victims of starvation and reach for their pockets to help the unfortunates. A common cry is why cannot we send our own surplus food to the starving? This is no solution except in the relatively few cases where a natural disaster has caused a temporary problem. A so-called drought, on analysis, is usually the normal dry time. In the past, with smaller populations, such countries turned to other drought-resistant resources; today, however, these resources have already been used to cater for the human increase.

What we need to send is not food but experts who will show every country in the world how to develop its own conservation strategy. Our foreign aid should be examined to see how it fits into the wcs guidelines. Then, instead of giving aid which often is in our own interests, we will send the help that will build a secure future for the developing country. Perhaps our greatest help has been to offer educational facilities to developing countries so that they can learn how to run their own affairs along wiser lines. Eventually such a policy will enable them to rely less on those offering aid.

Australia faces different problems. Our population difficulty is too many people in too few places. Eighty-five per cent of Australians live in urban regions, and cities such as Sydney and Melbourne are growing to such a size that they are becoming unpleasant and unmanageable.

Both historical and contemporary examples illustrate clearly that cities of a million people or even fewer can offer everything any reasonable human could desire. Consider the ancient Greek city-states with only tens of thousands citizens which achieved standards of living in artistic and democratic terms that are still the admiration of the world. Admittedly they had slaves to do the menial work; today mechanical slaves fulfil the same role.

Economists have shown how a *laissez-faire* policy allows cities to expand simply because the jobs are there, not because people want to live in large agglomerations. Every new resident increases the private affluence of land and factory owners. The public pays for the extra costs.

A more restrained world does not mean that we will lead duller lives in the future. Already in the Western world the service industries are showing the greatest expansion. Here people cater for the quality of life. We need better social services, particularly for the aged and disadvantaged groups in society. We need more scientists, artists and those who provide an increase in brain power and skills rather than consumer goods. In a manageable city, the size of Adelaide, the entire city centre can be covered on foot. In a thousand ways, small units and small cities are much more enjoyable than large ones.

Those who have visited the slums of the great cities of the world know that such places are an obscenity and only a small proportion of citizens enjoy the good life a planned city can offer.

Salting

We now realise that the unwise clearing of forests has led to a rise in a saline water table, turning productive farms into wastelands. In south-western Australia controls have been put on such clearing in water catchment areas; similar controls plus an intensive planting restoration campaign are needed in other parts of Australia where salting is a problem. The map indicates the extent and range of salting on the mainland.

Some problems do not begin and end with state boundaries—the Murray–Darling system has already been mentioned as an example. This has now been recognised and the lessons learned in the past should be put to good effect in solving the difficulties. It is not only a question of saving rural industries in one of the most fertile regions of Australia but is also vital to the supplies of fresh water to South Australia as the Murray provides much of this resource.

Other unwise uses of land

Chris Pratten, a one-time grazier and now environmental director of the New South Wales National Trust, highlighted a problem when he wrote an article attacking hobby farms. He was talking of smallholders rather than the kind usually stigmatised as Pitt Street/Collins Street/St Georges Terrace farmers, depending on the city in question.

A person who is successful in business and pours some of the profits into a run-down farm and by better management and capital makes it more productive is doing good. However, hobby farms are usually too small to be viable. Many are promoted by developers who take good farmland and sell small blocks of it as holiday homes or bush retreats. Most of these are around cities or along the coastlines where rainfall is good and soils are productive. Another kind of developer buys natural bushland cheaply and subdivides it into farms. Such land, being marginal, should be kept for its natural values. It can provide recreation for urban dwellers, but only in public terms.

The excellent report on the National Estate published by the Whitlam government had this to say:

> The forcing of farms onto previously unused or lightly used land has many costs and disbenefits. Land like this needs much more intensive use of machinery, irrigation equipment and agricultural chemicals to remain productive. It is also the land which until recently supported much of the original plant and animal associations. Some, at least, is unsuitable for intensive agriculture because of soil fertility, slope factors, or because the land may have other important values, eg catchment protection, recreation, wildlife. Greater returns for investment in agriculture and pastoral development are obtained from enhancing and increasing the yield from already developed land of known and improved productivity.

What applies to big farms applies with even more force to hobby farms. Chris Pratten pointed out that:

> Marginal lands are more prone to damage from soil erosion; noxious weeds and feral animal problems are greater; there is often greater fire risk, the cost of provision and maintenance of services is much greater as most of these areas are further from major centres of population and are often in rough and steep country.
>
> Exploitation of marginal land has also caused many problems which are an enormous cost to the community. Soil erosion, pasture deterioration, higher flood levels, spread of introduced weeds, silting of streams, rivers and water storages can result. Though the costs may not appear in ordinary accounting, they are finally and inevitably a burden on the whole community.

What is the solution? Obviously land speculators must be curbed. Those who do wish to have hobby farms must accept that they will pay for the total cost of all 'urban' services they manage to gain for their own comfort. Above all there must be a zoning system so that all good agricultural land is kept in production. Before any subdivision there must be a proper environmental study to ensure that any change is in the interests of the whole community, present and future.

A prime example of how changes in land use should not be done was the attempted army takeover of land in the Orange–Bathurst region near Sydney and the Cobar region further to the west The lack of proper planning which resulted in this land grab caused farmers and conservationists, traditionally not the best of friends, to join in a well-orchestrated protest which culminated in a Senate inquiry. The evidence showed the government that the project was unwise. Had this inquiry taken place before the army made its plans a great deal of money, and public disquiet, would have been saved.

This was a telling example of how the National Conservation Strategy was used to convince the government that no productive land should be destroyed unless there was no prudent or feasible alternative. It was an even more telling example of how conservationists can save the community money—hundreds of millions of dollars in this particular case.

The Bungle Bungles. This mountainous area was almost unknown to most Australians until about ten years ago. Now it is a tourist attraction, though good management will be needed to avoid this reserve 'being loved to death' by the tramp of tourist feet. (See also pp. 65–72)

Cradle Mountain chalet with rednecked wallabies in the foreground. The first chalet was built by Gustav Weindorfer who was one of the early conservationists lobbying for this region to become a national park. (See also pp 65–9)

4. Removing waste

Today a seemingly never-ending mass of rubbish threatens not only public land but also rivers and seas. Wetlands disappear, being ideal places (from some points of view) for landfill that turns mosquito-ridden swamps into playing fields or factory sites. Rarely seen dangers threaten rivers and lakes since much of the waste is soluble and remains hidden. Humans have used the oceans as one vast sink and poured into them the waste from a profligate society.

Today nature is sending out warnings that even the largest of sinks has a finite capacity. Poisonous mercury dumped into the oceans in the past is being recycled by marine bacteria away from their homes into ours by way of sea creatures such as school sharks and other predators. These gradually store an increasingly lethal dose in their bodies until they become unfit for human consumption and their catching is banned.

After a meal of oysters in Launceston some years ago I spent a disastrous night with my stomach galvanised into unwanted activity. Later I learned that water pollution was so high that the molluscs had 10 per cent by dry weight of zinc in their bodies, a byproduct from a smelting works nearby. I knew zinc was useful for galvanising iron to prevent it from rusting. I now know that it also galvanises human stomachs!

The zinc came from the Electrolytic Zinc Company which operated a refinery on the Derwent River. Since 1974 there has been a great improvement in the disposal of wastes. While the oyster farmers reached an out-of-court settlement, the company reduced pollution by developing a waste water recycling plant. This has reduced metal losses from the plant to 5 per cent of the 1973 levels. The money obtained from this recovery has offset much of the expenditure on new anti-pollution equipment. So reducing pollution can increase profits!

An even more recent example occurred when Greenpeace asked that the Hobart Electrolytic Zinc Company stop dumping heavy metals waste off the south-east coast of Tasmania. There were signs that seabirds may have been contaminated. This method of waste disposal had been permitted by the federal Department of the Environment in 1973. Since the waste was disappearing into water depths of 2000 metres, the technique appeared safe and is used worldwide. Scientific evidence now indicates that the practice is not as safe as once thought.

Recently it has been discovered that a dainty little seabird known as the fairy prion has cadmium levels of 14.5 parts per million. As yet this is not a dangerous level. Apparently the source is krill, crustaceans which feed on smaller life and in turn are eaten by larger forms including the largest of all the world's animals, the baleen whales.

Here is a classic example of how the small creatures of the seas are removing poisons even from the ocean depths and recycling them into surface food chains. Once again we have a warning of 'silent spring', this time among birds of the sea. The fairy prion may be like the canaries taken down coal mines in the last century to give warning of the dangerous gases in the tunnels. The canary, being more sensitive to danger, warned the coalminers it was time to leave. Is the fairy prion sending us the same message?

Stormwater drains wash fertiliser, pesticides, oil and other toxic wastes into rivers, lakes and the sea. (Douglass Baglin)

An exciting thriller of some years ago, *The Kraken Wakes*, told how an alien lifeform arrived on our planet and built an undersea empire from which it threatened human civilisation. How ironic if the fictional kraken was already in the deeps in the form of microscopic creatures that are returning our poisons, and in the process may help enfeeble or destroy the human lifeforms which threaten their placid undersea existence!

Choice magazine for June 1986 had a detailed article on the subject of toxic wastes and it made horrifying reading. To take only one example, Senator Colin Mason drew the attention of federal parliament to 160 tonnes of dioxin stored at the Union Carbide plant in Rhodes, a suburb of Sydney. Dioxin was the chemical involved in the Seveso disaster in Italy in 1976. He said that Union Carbide had disposed of other waste dioxin in dumps at Concord, Homebush Bay and Menai between 1949 and 1974. *Choice* described how scientists had told them 'off the record' where drums of wastes had been dumped into the ocean off Sydney, radioactive wastes poured down a laboratory sink, and other toxic waste tipped into sewers.

A Senate enquiry was 'appalled at the lack of accurate information' on this topic. It was estimated that in 1981, New South Wales and Victoria were generating 1700 tonnes of toxic metals and metal compounds annually and that New South Wales alone was generating 25 000 tonnes of strongly acidic and caustic material each year.

The question of wastes is not only a matter of health or aesthetics when litter pollutes cities and the countryside. It is also a problem of wasting resources of land and water. If minerals and energy are wasted, it is a matter of sheer survival. Waste of resources not only decreases the quality of our living but closes off options for a comfortable future.

To cover such a complex issue would need an entire book. Here I can only highlight a few problems, stressing that they are only the tip of a huge 'wasteberg'. Some general rules must guide us in all future planning. Better management methods will avoid some waste production, and recycling of waste products should be given much greater priority in research.

What are these wastes?
Organic chemicals
The worst of these are the polychlorinated biphenyls or PCBS. These are found in a number of substances including electrical equipment, paints, inks and hydraulic oils. They are hard to destroy and very stable so they travel along food chains. They are also highly toxic. It was thought that only high temperature incineration will destroy them but recent research indicates there may be a simpler and safer chemical solution.

Metals
These include cadmium, chromium, copper, mercury and lead, all of which may end in food chains. Being elements, they cannot be destroyed.

Oil pollution in rivers, estuaries and the sea
As the table below indicates, it comes as a shock to find where most oil spills start.

Source of oil spill	Percentage
Natural (from the earth)	12
Offshore wells	1
Fallout from the air	11
Rubbish dumps	5
Industry wastes	3
Urban road runoff	5
Wastes dumped in rivers	24
Tankers with load on top	3
Ordinary tankers	10
Tanker accidents	6
Cleaning bilges and taking on oil	10
Accidents	2
Other minor spills	8

Radioactive wastes

Nuclear power stations had a surge in popularity until most nations realised the problems; not only the possibility of accidents, but of how to dispose of a nuclear power station that is no longer of any use. In Australia we have found it impossible to solve a simple problem of radioactive waste in the Sydney suburb of Hunters Hill. The site of an old radium-processing factory has now been buried under a layer of soil and the open space turned into a park, a makeshift solution.

Because of environmental problems, nuclear power stations are no longer competitive with coal- or oil-fired ones. Any organisation turning to nuclear power must face the potential problem of the compensation claims should some accident, similar to the one at Chernobyl in the Soviet Union, occur. In Western countries such compensation problems are saved by the company going bankrupt, which leaves the costs to the community!

How can such wastes be treated?

Landfill

This poses possible disaster when the material begins to move into the water table.

Storage

This is merely putting off the evil day and leaks can be hazardous.

Ocean dumping

The cadmium story shows that out of sight is not necessarily out of animal bodies.

Deep-well injection

This is the placing of the waste either in old mineshafts or in new deep wells drilled for disposal purposes. It has possibilities but needs careful examination.

Chemical destruction

This may be practical as our technology develops.

Incineration

This is the best solution at present but it has so far proved impossible to find a suitable site close enough to Sydney or Melbourne where most of the toxic wastes are created. Most spills take place during transportation so the shorter the distance the wastes are carried, the better. An incinerator ship, the *Vulcanus*, travels the world to burn such wastes. Three offshore burning operations only halved the toxic liquid wastes stockpiled in Australia.

There seems no reason why Australia could not build its own ship and service not only its own requirements but also those of New Zealand and South-East Asian countries. If someone does not offer this facility most nations will take the easy option of dumping everything into that great sink—the sea.

It seems likely that only a disaster similar to the one in Italy or the Bhopal tragedy in India will cause governments to act. The Indian tragedy killed 2347 people and permanently

How much intractable waste?

	Stored (tonnes)	Generation rate (tonnes per year)
New South Wales	7400	700
Queensland	114	128
Victoria	186	50
Western Australia	51	50
ACT	11	5
Northern Territory	9	0.6
Tasmania	Not known	5
South Australia	10	15
Total	7781	954

These figures for amounts of stable organochlorine waste were compiled by the Australian Environment Council from a survey in 1983, and updated for 1985. The figures are indicative, and underestimate the true amount, which is unknown. The generation rate will decline as replacements are found for compounds such as PCBs, and efforts towards recycling are intensified.

Source: Ecos, 1987.

injured 86 000; another 203 000 had hospital treatment. Surely we must not sacrifice thousands of Australian lives before we take the steps which are so obvious. Those who clamour for a reduction in taxes should realise that problems like toxic wastes end up in the 'too hard' and 'too expensive' basket and are left until tragedy happens.

Perhaps a solution may be found in a proposal from Western Australia. The government has decided to build a high temperature incinerator at Koolyanobbing, 460 kilometres east of Perth, to burn the 1000 tonnes of PCBs stored as toxic wastes. The plant will cost only $1.5 million, which seems a cheap solution to a dangerous problem, and will be dismantled and sold once it has finished its work. Why cannot the other states adopt a similar policy?

The CSIRO Division of Applied Organic Chemistry operates a high temperature incineration unit at Fishermens Bend in Victoria. Dr Peter Waites, arguing from the experience of this small unit, considers that it may be the way of the future. The local community has accepted this unit so a number of small on-site incinerators may prove acceptable to other communities. Coupled with these it might be possible to develop a mobile incinerator to take the unit to the wastes, rather than the other way round.

There is still a need for a rotary kiln incinerator to deal with solid wastes. Providing this and the smaller units are licensed and satisfy environmental impact statements, the community would accept such solutions.

The latest information to hand is exciting: research at the University of Sydney may offer a way out. This involves a chemical process reacting at a little higher than room temperature which changes dangerous wastes into harmless salts.

Litter

This is the most obvious kind of waste. *Choice* magazine, in a challenging article titled 'Who's rubbishing Australia?', made some telling points.

The packaging industry is a powerful lobby. In 1971 the president of the world packaging association, Mr F. R. Briggs, had the effrontery to blame litter on the buying public: 'We don't make the pollution. We make the packages on request from manufacturers... Then the messy public discards them all over the countryside.'

He glossed over the fact that the packaging industry has managed to convince most state governments to refrain from taking legislative steps to stop litter at the source. Only South Australia has had the courage to take action. Laws in that state help stem the use of non-returnable containers and encourage their return and refill by offering a monetary incentive. As a result international experts have praised South Australia as being the cleanest state in Australia and one of the cleanest places in the world. Throwaway cans and glass bottles, once the major component of litter, have dropped to only 0.1 per cent of cans and 0.8 per cent of glass. The higher the rate for refunds, the higher the return, which is only to be expected.

Returnable containers increase employment since labour is needed to collect, clean and refill the containers. This makes the environmental law more costly for the individual factory but what is the balance sheet for the community? A survey in 1979 indicated that if Australia switched to returnable containers the saving in waste management would be between $30 and $40 million annually. This is only one economic factor. There are others and it is the paradox of environmental problems that what is dear for a factory is cheap for the community. In the long run such laws make costs fall. Providing that the laws do not discriminate against the individual, the factory owner has no problems unless there is

---

The Conserver Society

To qualify as citizens of a conserver society, we must shift entrenched attitudes and thinking. We need to recognise that there is rarely such a thing as 'waste': rather there are materials that sometimes end up in the wrong place.

The transition has already begun. The European steel industry re-uses scrap metal with energy savings as high as 50 per cent; in the case of copper, 90 per cent; and of aluminium, 95 per cent. Recycling a glass container saves only 8 per cent; but in parts of the United States, a citizen buying a bottle of soda or beer now pays a deposit against return of the empty bottle. If all drinks containers in the United States were to be re-used, the annual savings would amount to 0.5 million tonnes of glass—plus almost 50 million barrels of oil used in production processes. In Japan, OPEC spurred an increase in recycling of raw materials from 16 per cent to 48 per cent in just five years ('this year's Toyota is last year's Ford'). In Norway, the price of a new car now includes a disposal-cost item of about $100, redeemable when the junked car is turned in at an approved receiving centre. Major new businesses are emerging to exploit waste chemicals and oil. The thrifty Chinese claim they re-use 2.5 million tonnes of scrap iron each year, and at least 1 million tonnes of waste paper.

(The GAIA Atlas of Planet Management)

competition with overseas suppliers. This does not apply in the container industry.

Public opinion in South Australia is strongly in favour of the new system. Similarly, polls in America indicate that 75 per cent of people favour reusable containers.

The packaging lobby stresses the value of recycling campaigns and certainly there has been commendable progress in terms of the aluminium and steel industries. Nevertheless, this is a more energy costly method than using the containers a number of times.

The new laws in South Australia came into force in 1973 and the following figures are of interest in this litter problem.

South Australian Litter, 1973–81

| | Litter composition (%) | |
Litter type	1973	1981
Glass	2.8	3.8
Metal (cans)	10.2	1.7
Plastics	8.7	16.3
Paper	61.0	57.3
Miscellaneous	17.3	3.4
Beer, soft drink bottles	12.1	4.0

Energy wastage

This is another hidden cost for which the community pays. The figures from the United States give some idea of energy used in packaging, a resource which is wasted if the package is used once, then thrown away.

An example may make the story clear. The family that throws away 300 aluminium cans each year also throws away the energy used in making these cans. This is equivalent to a three-week supply of petrol for the average motorist. It represents a significant waste of our resources.

The tragedy of the lobbying which prevents the use of returnable containers is that so many people who care for the environment are encouraged to take part in 'band aid' programmes with the stated intention of keeping our country beautiful. Such programmes do some good but unfortunately prevent more important measures being taken. Politicians love these campaigns as they can obtain wide media coverage making emotional statements about the need

Seabirds are not only killed by oil spills. This fairy penguin was trapped by one of the plastic rings used to carry packs of beer.

to stop littering. They also enable them to avoid having to make firm decisions—politicians who do nothing stay in power longer!

One stopgap method for at least cutting the rubbish flood would be to compel every service station to have a large rubbish container. This should be serviced several times a week free of charge by the municipality. Instead of drivers and their passengers littering our urban and country areas they would be encouraged to deposit rubbish in one place. Some years ago large bins placed at pulloff places were removed by the New South Wales Main Roads Department on the grounds that too many people were using them to dump household rubbish! They were encouraged to drop it more evenly along roadside verges!

Recycling

All is not gloom and there are encouraging signs. BHP produces about two million tonnes of blast-furnace slag and two million tonnes of steel slag each year in its iron and steel smelters. Another two and a half million tonnes of fly-ash are caught in smoke stacks by the electrostatic precipitators of our coal-burning power stations. Bauxite refining produces 4.5 million tonnes of red mud each year which is either stockpiled or allowed to run into estuaries or the sea.

Sixty per cent of these waste-producing industries are within 50 kilometres of our cities. This means that they are close to potential users once we have found how the waste can be reused. Technology will gradually solve the problems of reuse.

Blast furnace slag, mainly used for landfill, can also be used for road making, concrete, glass and insulating wool industries. In the United States some cities are finding the slag so valuable that there is a shortage.

Dust from quarries is a nuisance but it has been used for making industrial tiles. Even the red mud from bauxite refineries may be used to improve the clays used in brick making. A CSIRO Division of Building Research is carrying out active work in this field. Those who echo the parrot cry of 'we have too many public servants' should start thinking more carefully. The CSIRO is a part of the public service.

Glass

Figures from Victoria indicate that about 70 per cent of glass waste is recovered for later glass making and this pattern is repeated in other states.

Aluminium

Recycling a can uses only 5 per cent of the energy used in making the metal for the original can. This has resulted in an increasing recycling of used cans.

Paper

About one-third of all paper used is recycled and this amount is increasing. However recycling is limited as recycled paper can be used only in lower grades of paper. To create better grades would be costly in terms of energy use.

An adequate agricultural policy is needed. Instead of growing unsaleable wheat, milk and other products we should be increasing our tree farms. So long as this is not at the expense of native forests but uses already cleared land, conservationists will approve.

Recycling benefits

1. A saving which means that resources can be used elsewhere.
2. Demands for raw material will be cut so that resources are not wasted. Recycled paper easing pressure on our forests is one example.
3. Some wastes can be burned to supply energy, so conserving fossil fuels.
4. Recycling reduces the need for good land being used for rubbish fill.
5. Less waste means less pollution of our waterways and oceans.
6. Many raw materials come from other countries. Recycling reduces our dependence and also saves their resources.
7. Recycling creates employment, an urgent need in developed societies where only a few days' work a week can produce all consumer needs.

Water waste

Water pollution is a major problem. Water storages can suffer a variety of pollutants including fertilisers of nitrates and phosphates which cause an outburst of plant life in dams. Other pollutants may percolate into the groundwater storages.

All cities need to control both agricultural and urban developments which may pollute catchment regions and intake areas of underground water. There is also a need to control private use of such underground reservoirs.

The major wastage of water is allowing it to run into the sea carrying with it a number of materials, the most abundant being human faecal matter. A recent article in *Ecos*, that excellent magazine which brings all kinds of biological and environmental facts to the general public, discussed the question of reusing this water.

Each day 2000 million litres flows out of our sewers into the sea. If we developed the technology for reuse we could use such water again and again for domestic purposes including drinking. After all, nature does it all the time, evaporating water from the sea and bringing it to earth as rain. It is an intriguing thought that the water in our own body may have begun its long travels through living organisms 3500 million years ago in early lifeforms. Nature never allows resources to become lost. She always recycles!

Such water, when the sewage has been removed, can be used to irrigate crops and pastures, recharge groundwater supplies and supply water for industrial purposes. The solid material in human faeces has been used for thousands of years in agriculture in many parts of the world. In hunter–gatherer societies all the waste material from animals and plants is recycled through the soil into new life.

The nitrates and phosphates which end up in our water supplies from farms and industries cause problems whether it is an outburst of plant life in the wrong place or difficulties on coral reefs in northern Australia. The CSIRO is working on the problem of how to make efficient use of this resource in Australia. Even if the refinements of making the water available for drinking prove too costly, economic treatment may make the water suitable for safe discharge into rivers and lakes or in irrigation for agriculture.

Agricultural use

Paris grows one-third of the vegetables it needs on sewage farms near the city. Nearer home at Werribee near Melbourne cattle and sheep have thrived on sewage irrigated pastures for 70 years. Over inland Australia similar techniques have assisted agricultural use. Experiments are being tried using sewage to fertilize tree farms to increase growth rates.

Even the problems of heavy metals in sewage may not be a major obstacle judging by results both in America and here in Australia. At Werribee, for example, dangerous metallic materials are held in the soil and do not travel into the plants which grow on these pastures. Monitoring will be needed to make sure this happy state continues.

Some optimistic scientists hazard that in the future we might discover a plant species which shows a preference for metals and take them up in large enough quantities to make it worthwhile extracting them. A kind of plant mine!

Aquifer recharging

Artesian and subartesian waters are a 'fossil' resource stored by nature often thousands or even millions of years ago. Like any store, they are not infinite. Recharging these aquifers is important and where nature does not provide enough rain it might be possible to use purified sewage water. Metallic substances would need to be removed but disease organisms such as viruses would die long before the waters were available for use by humans.

Industrial use

Obviously disease organisms would be no problem in the industrial use of recycled water, though metallic salts could be. Economic factors may make it too costly at present to pump treated water to where it is needed but in planning new cities this difficulty would not occur. The biggest problem is that most engineers are conservative. Opportunity occurred in the Gosford–Wyong growth region to develop such a sewage system. Conservation groups prepared a plan to reuse this water but it would have cost a few million dollars a year more than letting the sewage have an ocean outfall. Had the costs been measured in terms of both today and tomorrow the decision could have been different.

That is what a conservation strategy is all about. Taking the long view rather than the short view may save our children from expensive repair work. We have many historical examples of such disasters. Mine pollution at Captains Flat near Canberra is costing more in pollution repair than the value of the minerals in the past. Dozens of similar examples can be found both in Australia and abroad. Those who went before us cannot be blamed for the environmental sins they committed through ignorance or urgent need. We do not have that excuse. As trustees of the land we must consider our children's needs as well as our own.

5. Saving our wildlife

'Extinction is forever' is the slogan inscribed on the banner of the World Wildlife Fund. Some of my more cynical zoologist friends retort 'So what! Extinction has been a continuing process for thousands of millions of years. Look at the dinosaurs. It would be nice to be able to see a *Tryannosaurus rex* attacking a *Diplodocus* in a national park—the battle of the dinosaur titans—but we can survive without it.'

At the lowest level a world without its present diversity would be a duller world. Survival is not everything. Equally important is quality of living. More knowledgeable friends point out that a study of the fossil record shows that six major catastrophes have overwhelmed the wildlife of this earth, though the actual numbers of species were far fewer than we have today. It is amazing to realise that about three-quarters of all the then living species died out during the Permian extinction about 200 million years ago. The remaining wildlife of the time pulled out of such disasters so why worry today?

Saving wildlife—a world view

1. The preservation of genetic diversity is both a matter of insurance and investment—necessary to sustain and improve agricultural, forestry and fisheries production, to keep open future options, as a buffer against harmful environmental change, and as the raw material for much scientific and industrial innovation—and a matter of moral principle.
2. The issue of moral principle relates particularly to species extinction, and may be stated as follows. Human beings have become a major evolutionary force. While lacking the knowledge to control the biosphere, we have the power to change it radically. We are morally obliged—to our descendants and to other creatures—to act prudently. Since our capacity to alter the course of evolution does not make us any the less subject to it, wisdom also dictates that we be prudent. We cannot predict what species may become useful to us. Indeed we may learn that many species that seem dispensable are capable of providing important products, such as pharmaceuticals, or are vital parts of life-support systems on which we depend. For reasons of ethics and self-interest, therefore, we should not knowingly cause the extinction of a species.

(World Conservation Strategy)

It helps to know that it took 20 million years to come out of the Permian disaster. Twenty million years is not overwhelming to a geologist but to a human community it is forever. I had an excellent example of this in a public controversy with a distinguished geologist who also fancied his skills as a conservationist. We were arguing about a coral reef which had been destroyed by a cyclone and I had to admit that it had returned to something of its early interest.

'There you are,' he exclaimed, 'so if oil spills destroyed a reef it would come back.'

'It did,' I replied, 'but it took seventy years!'

For most of us it is little consolation to know that our great-grandchildren might be able to see what we had almost destroyed through ignorance or greed.

With extinction of species it is not a mere seventy years but millions. Also, earlier extinctions took place over millions of years, giving the survivors time to adjust. In contrast, modern extinctions are largely created by humans and at a frighteningly rapid rate, four hundred times faster, according to the experts.

Dr Norman Myers, a world authority on conservation who has publicised these findings, points out that human-caused extinction is worldwide. Rainforests cover 7 per cent of the world's surface and are under assault in South America, Africa and Asia. They hold about half the world's species of plants and animals and are being destroyed by logging and agriculture. A third have already gone. Seventy per cent of the losses are caused by increasing populations needing more and more of the forests. Given time to recover, their 'slash and burn' agriculture did not matter.

Conservation status of threatened Australian plants (number of species)

Conservation status	Uncertain	Distribution		Total	Percentage
		Very restricted (a)	Less restricted (b)		
Presumed extinct	41	26	11	78	3.5
Endangered	4	137	59	200	9.1
Vulnerable	1	335	276	612	27.7
Rare	2	441	409	852	38.6
Thought to be threatened	98	166	200	464	21.0
Total threatened	146	1 105	955	2 206	
Percentage of total	6.6	50.1	43.3		

(a) Species of very restricted distribution in Australia, and with a maximum geographic range of less than 100 kilometres. These species are in general at greater risk from localised threats (such as fire) than species spread over a greater range.

(b) Species with range of over 100 kilometres in Australia, but occurring only in small populations which are mainly restricted to highly specific habitats. These are subject to localised threats (such as drought, grazing, pests or diseases and land clearing).

Source: Leigh, J., et al., *Rare or Threatened Australian Plants*, ANP & WS Special Publication No. 7, Canberra, 1981, and Department of Arts, Heritage and Environment.

Today, when medical science has increased life expectancy, and religion or lack of social services prevents birth control measures, population pressure is creating disaster in the rainforests since there is no fallow period to allow the cleared forest to recover. The other 30 per cent of clearing is due to the 'hamburger connection' we have already discussed.

Dr Myers estimates that we could lose two million of our present five million wildlife species by the middle of the next century. In this book the term 'wildlife' covers both plants and animals. Coral reefs cover 570 square kilometres of ocean and are under attack in many tropical areas. They have a similar wealth of species as that of the rainforests. Wetlands are in demand for landfill to create more agricultural land or urban space for housing or factories. Australia has lost about 60 per cent of its coastal wetlands through these causes.

So there it is. A world of humans hurrying like lemmings to disaster at four hundred times the speed with which nature changed species through evolution.

What consequence has this for our future? Dr Myers points out that this new wave of extinction sweeping most of the world means a drastic change in evolutionary patterns since nature will have a much smaller gene pool on which to work.

There are some hopeful signs. Huge marine parks such as that of the Great Barrier Reef indicate that Australia is leading in one field. The recent inclusion of New South Wales rainforests and the rainforests of north Queensland on the World Heritage List is another indication of progress. Taken as a whole Australia has a reasonable percentage of its land as national parks, with a weakness in Queensland, the state with the worst environmental record.

Yet national parks are merely islands in a sea of agriculture. It is well known that island populations are more vulnerable than those of continents. Many larger animals need more room to move than offered in the normal sized park. Elephants are an example, and some scientists claim that we will lose most of our larger land animals.

This problem emphasises the importance of wildlife corridors and is the reason why the Wild Life Preservation Society of Australia sponsored the idea of 'koala corridors' as part of the Greening Australia tree-planning campaign. What is good for koalas would be good for other wildlife. Such green 'roads' leading from reserve to reserve would form a network allowing free passage of wildlife all over Australia.

Dr Myers makes the following appeal: 'We can still save species by the millions. Should we not consider ourselves fortunate that we alone among generations are being given the chance to support the right to life of a large share of our fellow species and to safeguard the creative capacities of evolution itself?'

It is not always realised that in our plants and animals we have a national treasure, something we should safeguard for the world. Few of us know that the eucalypts are among the most valuable of all world trees, providing timber, shelter belts and a number of other values which have made them important for foresters around the world. Two other groups even less well known are the sheoaks and the wattles. Our big three are still growing in value with at least seventy nations using some of them.

The NCSA sums up what needs to be done as a priority national requirement in the following objective: 'to preserve the genetic diversity of Australia's plant and animal species and ecosystems and of those introduced species which support plant and animal based industries.'

Grants for koala research

Financial grants have been made to three Australian scientists for 1986–87 to continue research into koala biology, including the disease Chlamydiosis.

Chlamydiosis causes a range of symptoms in koalas including infertility, pneumonia, blindness and urinary tract infections, and threatens some koala populations.

Source: Ecofile, 1987.

How are we trying to keep genetic diversity and how successful have we been? Are we planning to do better in the future? What have we lost? Most Australians know something of the extinction of our animals but few realise that our plants have suffered even more. Indeed all over the world the same pattern holds. It is possible that we will lose 10 per cent of all our plant species unless action is taken now.

Even more important than single species is the conservation of groups of plants or 'alliances'. Large groupings include areas such as rainforests but the figures here deal with smaller categories and help tell the story of urgent need in some parts of Australia.

Conservation status of alliances recorded in each state of Australia

State or territory	No. of alliances recorded	Percentage of alliances conserved		
		Excellent	*Reasonable/Moderate*	*Poor/Nil*
New South Wales	202	13	42	45
Northern Territory	77	13	36	51
Queensland	153	<1	22	78
South Australia	127	2	58	40
Tasmania	305	28	47	25
Victoria	125	2	41	57
Western Australia	218	2	49	49

Source: Specht, R. L., Roe, E. M., and Boughton, V. H., *Conservation of major plant communities in Australia and Papua New Guinea,* CSIRO, 1974.

We do not know how many animal species the coming of the Aborigines doomed to extinction. It is likely that both direct hunting and the increased use of fire destroyed a number of species, particularly the larger marsupials such as the rhinoceros-sized diprotodon, the giant kangaroo, the giant goanna and the marsupial lion. In more recent times, perhaps only 3000 years ago, the arrival of the dingo, probably a camp follower of a new human invasion, must have caused as much havoc among ground-dwelling animals as the

Right: *Koala mothers and babies now have a safer future. Sixty years ago they were slaughtered in millions for the fur trade. The present dangers are loss of habitat and frequency of bushfires.*

fox and feral cat did in the last two hundred years. It is almost certain to have been the major cause of the disappearance of the Tasmanian tiger and devil from the mainland, although they survived in Tasmania where the dingo never reached.

This is the official list of mammals and birds thought to be extinct since white settlement. It is taken from Derrick Ovington's *Australian Endangered Species*. The official list of species in danger is much larger.

Some species probably extinct following European settlement

Species	Date of extinction
Lesser stick-nest rat *Leporillus apicalis*	1933
Darling Downs hopping mouse *Notomys mordax*	c. 1840
Big-eared hopping mouse *Notomys megalotis*	1843
Tasmanian emu *Dromaius novaehollandiae* subspecies	c. 1873
Broad-faced potoroo *Potorous platyops*	c. 1875
Greater hare wallaby *Lagorchestes leporides*	1890
Night parrot *Geopsittacus occidentalis*	1912
Paradise parrot *Psephotus pulcherrimus*	1922
Pig-footed bandicoot *Chaeropus ecaudatus*	c. 1926
Toolache wallaby *Macropus greyi*	1927
Thylacine *Thylacinus cynocephalus*	1936

Source: *Australian Endangered Species*.

A roadside scene in South Australia. The first farmers regarded trees in particular and the bush in general as enemies to be destroyed. Luckily, plant corridors were left along roadsides. Now the government is working to restore some of the earlier plant life in a 'greening' of the State. It has been found that this is not only good for wildlife but also for the farmers. (See also pp. 68–70) (SA Government photograph)

The Basin, Rottnest Island, near Perth, Western Australia. Most Australians travel to the coast for their recreation. Rottnest Island is a reserve and tens of thousands of tourists flock here during the summer with some coming at all times of the year. New South Wales has set aside a third of its coastline as national parks. (See also pp. 72–3)

Conservation status of threatened Australian non-marine vertebrates (number of species)

Non-marine vertebrates	Conservation status				
	Presumed extinct	Endangered	Vulnerable	Rare	Total
Mammals	17	9	19	–	45
Birds	1	9(a)	10	2	22
Reptiles	–	5	4	1	10
Amphibians	–	2	6	–	8
Fishes	–	3	1	–	4
Total	18	28	40	3	89

(a) Includes four species restricted to Norfolk, Christmas or Lord Howe Islands. Species presumed extinct from these islands are not included.

Source: Adapted from Ride, W. D. L., and Wilson, G. R., 'The Conservation Status of Australian Animals' in Groves, R. H., and Ride, W. D. L., eds, *Species at Risk: Research in Australia*, Australian Academy of Science, Canberra, 1982, and Department of Arts, Heritage and Environment.

It is significant that the birds have managed to escape the disasters that overwhelmed many of the mammals. Perhaps the explanation is that most do not live on the ground all the time and can escape by flying. New enemies such as foxes and cats were supplemented by domestic stock and rabbits. Sheep and rabbits not only competed for food but also removed the long grass and tussocks that were the shelters of so many native mammals.

What is to be done? Crying over spilled genetic resources serves little purpose beyond learning what mistakes to avoid in the future. What we must do is save what's left.

Reserves

There are many kinds of reserves in Australia, their names varying from state to state. The best known are the national parks, relatively large areas set aside because of features such as unspoiled natural landscapes with their associated wildlife. These areas are dedicated for public enjoyment, education and inspiration and protected from all forms of interference except that needed for essential management. If an area is 'permanently dedicated', only parliament can destroy it. In the long run this is the only protection any democracy can offer a reserve. Backing that legislative barrier is the force of public opinion.

The strength of public opinion was shown clearly when the Queensland government planned to sell a Great Barrier Reef national park to a private developer. Backbench members of the National Party revolted against this desecration, indicating how deeply imbued is the national park philosophy in even the most conservative of political parties, despite the fact that such feelings were not shared by their executive.

The last few years have seen the worldwide development of wilderness reserves. True wilderness is a region where human activities are not obvious. Visitors must explore without mechanical aids. In so doing they gain a new vision of nature and of themselves.

Nature conservation reserves in Australia, 31 December 1986

Area of nature conservation reserves

The area of Australia excluding external territories is 768 242 785 hectares, of which 34 530 702 hectares or 4.49 per cent of land area is reserved for nature conservation as at 31 December 1986. The distribution of such areas within the states and territories are listed below. Terrestrial reserves of the external territories are also included in this list.

	Number	*Area (hectares)*
Australian Capital Territory (total area)		**240 000**
National parks	1	94 000
Nature reserves	2	10 020
Reserves	3	7 821
Total area of reserves		111 841
Land reserved for nature conservation		46.60%
New South Wales (total area)		**80 160 000**
National parks	66	2 892 259
Nature reserves	180	511 513
State recreation areas	20	20 749
Aboriginal areas	9	11 519
Historic sites	15	2 899
Total area of reserves		3 438 939
Land reserved for nature conservation		4.3%
Northern Territory (total area)		**134 620 000**
National parks	5	321 513
Aboriginal national parks	1	220 700
Australian government national parks	2	1 439 931
Conservation reserves	13	46 768
Nature parks	20	24 803
Game reserves	1	1 605
Historical reserves	13	5 874
Other conservation areas	31	1 718 049
Total area of reserves		3 779 243
Land reserved for nature conservation		2.81%
Western Australia (total area)		**252 550 000**
National parks	55	4 445 272
Reserves	8	1 083
Nature reserves	1 261	10 203 436
Total area of reserves		14 648 708
Land reserved for nature conservation		5.8%

Queensland (total area)		**172 720 000**
National parks	304	3 376 898
Environmental parks	141	44 271
Fauna reserves	2	25 906
Fauna refuges	5	5 873
Scientific purpose reserves	6	39 083
Total area of reserves		3 492 031
Land reserved for nature conservation		2.02%
South Australia (total area)		**98 400 000**
National parks	12	2 634 123
Conservation parks	184	4 049 831
Recreation parks	14	4 522
Game reserves	10	22 494
Total area of reserves		6 710 970
Land reserved for nature conservation		6.82%
Tasmania (total area)		**6 792 785**
National parks	13	851 140
State reserves	54	19 965
Nature reserves	38	29 426
Game reserves	8	2 770
Conservation areas	32	33 161
Muttonbird reserves	5	9 288
Aboriginal sites	4	1 243
Historic sites	30	793
Total area of reserves		947 786
Land reserved for nature conservation		13.95%
Victoria (total area)		**22 760 000**
National parks	30	960 311
Other parks	34	325 346
Other parks and reserves	9	5 839
State game reserves	66	38 979
State nature reserves	87	70 709
Total area of reserves		1 401 184
Land reserved for nature conservation		6.61%
External Territories		
National parks	2	**2 830**
National nature reserves	3	327
Total area of reserves		3 157

Source: Australian National Parks and Wildlife Service.

Regional parks

Day breaks. . .
Sun rising, scattering the darkness; lighting the land. . .
She shines on the blossoming coolibah tree,
Shady branches spreading. . .

These lines, from part of a longer poem of the Mudbara tribe of the Victoria River country in the Northern Territory, as translated by Ron and Catherine Berndt, indicate the aesthetic pleasures of landscape enjoyed by a hunter–gatherer people, the first Australians. Their enjoyment was shared by other non-literate people around the world, the best known examples being the stories of the Red Indians of North America.

With the rise of agriculture, the enjoyment of landscape and privacy became the prerogative of the rich. William the Conqueror, in a passion for hunting, ruthlessly recreated a wilderness forest out of farmland; ironically it was called the New Forest.

As the years passed the powerful were able to surround their homes with pleasant, ordered, open spaces. As rulers they extolled the virtues of warm, city life with numerous human contacts, but in their own lives they demanded space. In England, learning from Chinese and Japanese exponents, the landscape gardeners they employed were able to produce what appeared to be a natural landscape.

Yet some hankered for more. The American philosopher Thoreau epitomised this need in the phrase: 'In wildness is the preservation of the world'. The declaration of the world's first national park, the famed Yellowstone, was made in 1872. Australia echoed this lead with the declaration of the Royal National Park south of Sydney in 1879.

National parks best satisfy the need stressed in the World Conservation Strategy regarding genetic diversity. Adequate size is important to keep them viable and the naive idea that one can pick bits and pieces as parks and use the rest for a variety of other uses is scientifically laughable.

Yet even when we have at least 5 per cent of our country reserved in this way we still have the other 95 per cent to consider. National parks are the crown jewels of our wildlife heritage, but they are strung on the golden thread of small reserves, wildlife refuges, roadside edges, home gardens, ornamental lakes and other places were some wildlife survives.

Reserves are all islands in the vastness of Australian land and seas. We hope national parks will reserve for all time the diversity of the Australian landscape, but we also need smaller reserves within easy reach of everyone, places we know intimately and in which we feel rested, comfortable and at home. A single tree can provide a place to be born, a place to make love and finally a place to die for a cicada.

Another planning concept is slowly being realised in Australia. This is the regional park. The Great Barrier Reef Marine Park, the largest marine reserve in the world, is a magnificent example of the concept.

The concept of the regional park arose in England, though unfortunately the government called them national parks, in order not to be left behind by the progressive Americans! This self-deceiving name has led to a lot of confusion. First created in the beautiful Lakes District, such parks contain not only natural landscapes but also farms, villages and cities.

The more logical French call these controlled areas *parcs naturels regionnaux*. Translating this into regional parks is perhaps the best compromise. They are large areas intended to provide recreation in the best sense for an urban people, and which allow traditional economic activities to continue in the framework of a management plan. Public access is provided over private land with compensation to landowners for any loss they may suffer from making their fields open to walkers.

In France the initiative for setting aside regional parks comes from the local authorities. A management charter which has to be approved by the local authorities, the regional authorities, and the state is developed.

Some Australian states have been preparing lists of such potential regional parks while National Trusts, particularly in New South Wales, are engaged in classifying such areas in similar fashion to the way they classify buildings which need to be preserved. For example, the Queensland Conservation Council suggests the following as potential regional parks: sections of the Darling Downs, Tamborine Mountain, farmland around the Glasshouse Mountains, parts of the Atherton Tableland, Bribie Island, Blackall Ranges and various river systems.

Every state has its favoured cultural landscapes—the Hunter River region in New South Wales; Mornington Peninsula in Victoria; the Derwent Valley in Tasmania; the coastal strip from Cape Naturaliste to Cape Leschenault in Western Australia; the central desert regions around Alice Springs in the Northern Territory, Kangaroo Island, the Coorong and the Flinders Ranges in South Australia. For those who worry that intense public use will destroy the beauty to be preserved there is in Holland a tiny reserve of a few hectares of coastal dunes which attracts 700 000 visitors a year. They do no damage because of skilled management policies. Massive use still allows the preservation of areas of wildflowers and lakes with abundant wildlife.

Author Louis Bromfield, in the *Audubon* magazine for 1963, summed up the ideals of regional parks: 'Instead of a wasted and ruined countryside, crossed by polluted streams, devoid of wildlife and beauty, it is possible to make of the whole nation a vast and wonderful park in the midst of which works humanity, surrounded by a kind of natural paradise. All this is far less of a dream than it might appear; it is merely commonsense. More than that, it is profitable not only in terms of dollars and cents but in a thousand other ways . . .

The pattern is merely that of humanity working with nature and creating the environment which is our proper birthright . . . if we choose to claim it . . . '

When we reach a conservation Utopia, Australia will have a web of national parks sampling all our rich diversity of scenery and wildlife. Surrounding these crown jewels will be a setting of regional parks so that our whole land and sea will be under careful and loving management.

It is only commonsense to work with nature and live in beautiful, clean and interesting surroundings. In *Look Here*, published in 1968, a number of authors tried to develop a greenprint for the Australian environment, natural and human. I began by quoting Sir Macfarlane Burnet:

If we obtain pleasure from watching and understanding the birds and beasts of our countryside, we have a clear obligation to leave a similar opportunity for future generations. Who would want to live in an overpopulated world, where outside the urban sprawls there are only vast agricultural factories in the field?

and I concluded:

Is it too Utopian to believe the basic principles that control the world's regional parks could not be used to control all developments throughout the nation? Why not an Australia in the year 3000 AD with cities of half a million people surrounded by green belts of farms, national and regional parks? These would provide all types of Australian environments where people could move easily from the cities for the refreshment and recreation they need.

I have not changed my mind in the last twenty years!

National Heritage Parks

The World Heritage Convention links the world's cultural and natural heritage and makes people aware of both its value and the grave dangers it faces. It is an attempt to encourage every nation to accept responsibility for its own heritage. The World Heritage List is a prestigious one and nominations are not accepted lightly.

Australia already has seven items, a considerable number compared with the rest of the world. We must accept with good grace the rejection of some of our future nominations. The national parks of the Blue Mountains of New South Wales, the Green Mountains of Queensland, Wilsons Promontory in Victoria, Flinders Chase on Kangaroo Island, the Stirlings and the Bungle Bungle in Western Australia, are all areas of great natural value. Similarly the Sydney Opera House, Port Arthur and Kingston on Norfolk Island are places of great cultural value. But would they be accepted for the World Heritage List? Possibly, but not certainly. How then can they best be protected?

The Wild Life Preservation Society of Australia has put forward a concept to bridge the gap between items of World Heritage and National Heritage. Although irreplaceable, these cannot be considered for the world list without making that list so huge as to lose significance. Natural areas might be called National Heritage Parks and places of cultural significance might be called National Heritage Sites.

The society is concerned in particular with National Heritage Parks. It suggests that, by mutual agreement, state governments should negotiate with the federal government on particular parks they consider of national significance. In return for federal assistance, both financial and through the expertise of the Australian National Parks and Wildlife Service, a state would accept that management plans for such a park would be developed co-operatively. A condition of continuing federal financial assistance would be that no change in management plans which would alter the character of a place would take place without mutual agreement. In this way the ownership of the land would remain as it is today but management planning and financial responsibility would be a joint affair.

'Small is beautiful' applies to many fields and the Wild Life Preservation Society does not suggest an all-embracing Australian National Parks and Wildlife Service running all Australia's national parks. The establishment of National Heritage Parks would be a

Australian Alps national parks

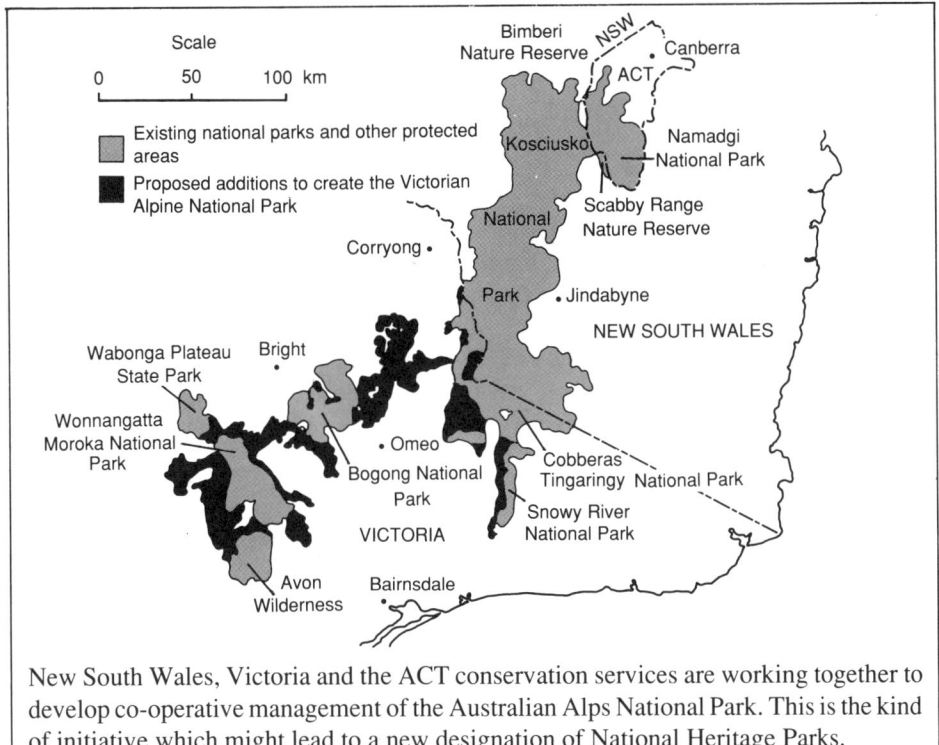

New South Wales, Victoria and the ACT conservation services are working together to develop co-operative management of the Australian Alps National Park. This is the kind of initiative which might lead to a new designation of National Heritage Parks.

Source: Australian National Parks and Wildlife Service.

method by which those states unable to provide the money for satisfactory management of present parks would receive help.

The states could also make a survey to determine other areas that need national park protection. This is already happening on a small scale. The Australian government is assisting any state government which needs funding and providing other assistance for particularly valuable state parks. Perhaps the most publicised instance, and a good omen for the future, was the decision on high country alpine parks. Victoria, New South Wales and the Australian Capital Territory have joined to develop management plans for the huge national park which is spread over the mountains of south-eastern Australia.

In this way state and federal governments would work together to keep those Australian 'lesser crown jewels' safe for all time.

Marine parks and reserves

One weakness in our present reserve system is in the sea. Until the formation of the Great Barrier Reef Marine Park little attention had been given to the conservation of seashores and reefs.

Control over these areas is often split between a number of organisations and state and federal governments. In 1986 the Australian committee for IUCN produced a report on our marine and estuarine areas with a policy for their protection. Incidently, the marine extension to the national park of Bouddi was the first marine national park in Australia.

An inventory of declared marine and estuary protected areas in our waters was published by the Australian National Parks and Wildlife Service. It stated: 'The first marine protected area in Australia was declared in 1938 at Green Island off the Queensland coast. Since then, nearly 37 million hectares have been afforded protection under various marine and estuarine protected (MEPA) categories. This represented less than 0.06 per cent of the total marine areas for which Australia is responsible'.

It is accepted that at least 5 per cent of a country should be reserved as national parks and similar reserves. These must be chosen carefully to sample the diversity of a country. It also means that anywhere in Australia a person should have reasonable access to at least one park. On that basis we still have a long way to go in terms of our marine resources.

Reservation

Figures change with time as governments add to reserves. Some, like Kakadu and Uluru, are national parks whose ownership has been transferred to Aboriginal groups who have leased them back to the government. Those regions that have passed into Aboriginal management will suffer little change so long as the old hunter–gatherer tradition continues, although non-owners will find it difficult to obtain permission to enter such areas. This is unfortunate as it has been a 200-year-old tradition on outback properties for travellers to be given free access and the right to camp near roads. Aboriginal groups, like other Australians, can be tempted by greed with potentially unfortunate environmental results.

Tourism

One concern among conservationists has been summed up in the American phrase that national parks can be 'loved to death' under the tramp of tourist feet. Attacks on tourism come not only from some conservationists but also from some scientists.

Specialist pleading must be treated with caution. Such folk never consider themselves as being tourists. This may be because they are being paid for visiting some of the most beautiful places in Australia. Of course their work is important but there is little evidence to justify the belief that they are more caring of the environment than tourists.

In my travels across central Australia I have found litter strewn over hundreds of metres by scientific survey groups. In Antarctica there have been horrifying stories of discarded material from research stations. More recently there have been cases of penguin colonies destroyed to create airstrips. The problems caused by a constant stream of anthropologists have been documented in many parts of the world.

My own experience has provided no evidence that tourists have caused serious problems in any reserve in Australia where there has been competent ranger management. The opposite often occurs. Pandas, tigers, rhinoceros, koalas, woodhens and a myriad other species would be extinct except for the fact that enough people were enthralled by them and were determined they should not disappear.

> ### *Wildlife as a tourist resource*
>
> ... wildlife is a major resource base for recreation and tourism. Tourism, largely based on wildlife, is among Kenya's top three foreign exchange earners. In Canada 11 per cent of the population hold hunting licences; in the United States 8 per cent hold hunting licences and 13 per cent hold fishing licences; and in Sweden from 12 per cent to 18 per cent hold fishing licences. Many more people enjoy simply looking at wildlife; in the United States there are about 7 million birdwatchers, 4.5 million wildlife photographers and almost 27 million nature hikers. For a great many people, too, wildlife is of great symbolic, ritual and cultural importance, enriching their lives emotionally and spiritually.

(World Conservation Strategy)

National parks around the world have been the main tool of wildlife conservation. This would never have happened had it not been for the thrust of tourism, often spearheaded by the minority of scientists who cared enough to see beyond their own research. The tourist industry has been the cash crop which has inspired governments to decide to save the world's greatest natural attractions.

Heron Island provides a good example. It was a turtle canning factory until conservation pressure stopped this industry. Then tourism took over on an island where the black noddies were a feature though the colony was small. The tourists increased in numbers. So too did the noddies until today they are present in tens of thousands. Near Cairns is another island, Michaelmas Cay. Earlier this century the terns were harried by commercial egg collectors and vandals. Gradually more and more visitors to the cay complained about the interference to the noddies and sooty terns. Their views prevailed and today the birds nest year round. A simple request to all visitors to keep on one side of the island and walk only on the beach has meant the nesting birds watch placidly as the tourists enjoy themselves on the beach.

On the other side of the continent is Pelsart Island, once a tourist resort. Red-tailed tropicbirds came to nest and their numbers rose steadily with the strict protection offered by the hostel owners. The tourist camp closed and soon the tropicbirds were no more.

Visitors around the coast and inland are unofficial rangers. Fishermen everywhere with no scientific justification regard most seabirds as enemies. They use the larger kinds for cheap bait for spiny lobster pots. I have visited islands where cormorant rookeries have been destroyed by the heavy boots of the ignorant, trampling eggs and young.

The conservation lesson is simple. Conserve the habitat, lay down sensible management rules for visitors, and you can have your tourist cake and eat it too.

What do we have?

In a country as large as Australia the problem of mapping our wildlife resources is not simple. Many groups and individuals have added grains of new knowledge to the total. In the last few years, however, the earlier haphazard work has been co-ordinated with the formation of the Australian Biological Resources Study, set up as a permanent body by the Australian Department of Science.

The main task of this group is to co-ordinate taxonomic work at the national level. Taxonomy is the study of species. It is vital to know what a particular plant or animal actually is, whether it is part of our natural heritage, part of our economic resources or to be destroyed as a potential pest. Knowing species is the foundation stone of wildlife conservation. Another major task is to organise studies of those of our plants and animals that are receiving little attention. Bird study is popular while spiders, slugs, worms and a host of other creatures are not regarded as worthwhile by the average amateur naturalist. For example, we spend millions of dollars trying to solve the problem of managing our large kangaroos but very little on the endangered smaller members of the kangaroo family.

Management

Over Australia the position is like the proverbial curate's egg, good in parts. There has been a steady improvement during the last twenty years, although the last few years have seen a retreat due to lack of funding, forcing staff to manage much larger areas.

Tasmania, Victoria, South Australia and Western Australia have recently merged such services into wider government departments. Governments claim this merging will make them more efficient; conservationists have their doubts. Without direct access to the relevant Minister, life becomes difficult for reserve managers. A major problem is that so little tertiary training is available in research-directed management.

Yet management is not as important as setting aside the reserves. Nature has been managing such regions with reasonable success for thousands of years; the lack of human management should cause little permanent damage. A national park manager once asked a distinguished conservationist what action he should take to manage his park. 'Sit on your hands' was the answer. Judicious neglect is often the best policy as long as the total area is viable in terms of surviving on its own.

How large?

The ability to keep all the animals and plants of a region in good shape depends to an extent on how large the area is. One method of finding a figure is to examine islands as this should give an idea of whether the largest animals can survive in a given area.

National parks that have been in existence for many years should also provide evidence, though parks must be viable for thousands of years if they are to carry out the needs of the future. Studies indicate that the larger the area, the more species can survive. The closer it is to other reserves the more chance there will be movement of species from one to the other. A park with a circular boundary is better than one with a long thin boundary. In certain habitats, however, particularly estuaries, a long, thin park may be more efficient.

M. P. Bolton and R. L. Specht, in a study of western Queensland, devised a method for selecting conservation reserves by a quick assessment of the plants in a region. An area with the greatest diversity of plants is also likely to have the greatest diversity of animals.

In the past many of our national parks were selected because at the time the government had no other use for them. It is now the job of park planners to look for gaps in their system and fill them as soon as possible. Often this must be done secretly. The price may escalate as soon as owners realise that the government is interested in buying the land. Even more worrying is the 'blight' that settles on an area when it is indicated that at some time in the future a park will be declared with a compulsory acquisition. State governments do not have a good record in this field and owners can suffer financial loss because of it.

Future legislation

Conservation groups have spent a great deal of time and money working on the problem of the most effective legislation to ensure the protection of the wildlife treasures of Australia. At the core is the need to conserve habitats rather than concentrating on endangered species. Michael Kennedy and Ross Button have produced *A Threatened Species Conservation Strategy for Australia*. This booklet lists threatened species including mammals, birds, reptiles, fish, worms, molluscs and arthropods. Plants are also included. The authors discuss the necessary strategy components and the legislation needed, and include a model Endangered Species Habitat Act for New South Wales. It is obvious that in a federal system we need complementary legislation for all the states and the Australian government. Should one state make a beginning in this field, in time others will follow.

The gene pool

Every year sees new discoveries useful for humans in that vast source of genetic material to be found in our native bushland and the depths of the ocean. For example, a study of sea whips, a kind of coral, has indicated that they may provide important pain-killing drugs. The pharmaceutical industry is one of the main users of our native plants and animals. In the United States the commercial value of all medicines of natural origin now reaches ten billion dollars annually.

On the pesticide front biological control products in California alone have saved more than $200 million by reducing the need for chemicals. One estimate has shown that every dollar spent on research into biological controls results in a thirtyfold economic return.

Even though chemists are developing new pesticides (one recent discovery, after a twenty-year study by the CSIRO, is a breakthrough in treating such problems as the sheep blowfly, now marketed as 'baclash'), they are not as useful in the long run as using nature to fight nature!

Genetic engineering using our natural resources may bring huge dividends. Every species is a kind of biochemical factory we destroy at our peril. Penicillin, a simple fungus, has produced more benefits in terms of alleviating human suffering and death than any other material and yet most people see fungi as either useless or pests. Many of them are, but we should never destroy them unless there is an urgent and proven need.

6. Sustainable production

Farming

'In essence the widespread land degradation that has occurred in Australia since European settlement may be attributed to lack of knowledge by the land user, or disregard of long-term adverse effects for short-term advantages.'

In these words the Senate Standing Committee on Science, Technology and the Environment summed up the story of land use policy in Australia. In a more optimistic mood the committee believed that almost 200 years of ignorance and greed may be coming to an end, since from the 1970s we appeared to be passing into a new phase.

The report also highlighted the problems of a federal system where 'there is no single authority for land use. Land use policies are fragmented with inconsistencies between states, between agencies and departments, within states, and within departments.' It concluded that a national policy is needed, not only for the problems which are found over Australia but also for some regions recognised as being of national and international significance. Obvious areas are possible World Heritage sites, rainforests and the high country—all regions which are beginning to receive national attention.

To ensure that such policy statements produce action, the report suggested a Standing Committee on Land Use be established to monitor our land use so that we can analyse, interpret, draw policy conclusions, and then make sure these are co-ordinated and consolidated.

One warning has been given by conservationists which should be the basis of our future. We must develop the idea and ideal of trusteeship, an acceptance that we do not own the land in the same way as we own a motor car or an item of clothing. We are trustees only, inheriting land from those who went before us and obliged to hand it on to those who come after, in even better health. In other words we need to develop a land use ethic!

Yet before looking to the future we need to look at our past. What lessons are to be learnt from the last 200 years? We can look even further back to the thousands of years in which humans have been agriculturists, since their ideas were transplanted from the Old World to the New without any consideration of the fact that European land practices may not be suited to the soils and climate of Australia.

The humans who evolved millions of years ago would have foraged in similar fashion to other primates, gathering roots, seeds, fruits and other edible parts of plants. They obtained some material secondhand by eating other animals, from insects to creatures as big or bigger than themselves. Even a mammoth could be killed with a weapon!

Generation succeeded generation till a more innovative people noticed that soil which had been dug in search of edible roots, or harvested for seeds, showed increased productivity. It is interesting that although the Aborigines of northern Australia were familiar with the cultivation methods of the Islanders and the Macassans they never imitated them, being satisfied with what nature provided. Even so, the conventional wisdom that hunter-gatherers in general, and Aboriginal Australians in particular, pressed lightly on the

The Tasmanian tiger was probably exterminated on the mainland by competition with dingoes brought by Aborigines several thousands of years ago. It was probably exterminated in Tasmania by a shooting campaign because occasionally it killed settlers' sheep. The continuing destruction in this century was a stupid act of government and led to its extinction. Occasional supposed sightings keep alive the hope that a few may survive in the western wilderness areas.

country, needs some modification. Major changes began long before the arrival of white settlers.

With the coming of the Aborigines some 50 000 to 70 000 years ago, and possibly even earlier, the larger animals such as diprotodons, already suffering from changes in climate, would have had their final push into extinction. Modern Aborigines were able to kill the introduced water buffalo, so had the technology to destroy such marsupial giants, today found only as fossils. The coming of the dingo, perhaps about 3000 years ago, must have caused even more havoc.

The greatest change people brought to the land was the use of fire as both a hunting and agricultural tool; what today we call natural is in fact a human artefact. There is some evidence for a major change in plant life and increased erosion soon after the Aboriginal arrival. One example of this process is the gradual spreading of the rainforest near Brisbane over the last few decades, as can be seen in the Lamington National Park and at Mount

Glorious. In both places fire has been kept out. The reverse can be seen in the hills above Cairns where fire has caused the rainforest to retreat.

In my wanderings in sandy deserts to the north of the Nullarbor and across the vast regions of limestone to the wooded coastal fringe, I have often wondered if this almost treeless plain is natural. There is only a thin coating of soil but where this is deeper, in hollows, trees do grow. On Nullarbor station the owner planted some tuart trees, eucalypts which flourish on limestone. These have grown some 15 metres high. Was it then the use of fire which removed a fragile vegetation in the dry times of 20 to 15 thousand years ago, so that strong winds could shift the soil both north and south leaving a stony desert?

There is no question that the coming of the white settlers caused massive changes. With them came hard-hoofed domestic stock that broke the soil surface while their close-cropping teeth often proved disastrous to plant life which had only felt the touch of the more gentle marsupials. One agricultural economist put the problem succinctly by stating that we are the lucky country because we have been here only 200 years, not long enough to create the deserts produced in the Mediterranean regions by thousands of years of grazing by sheep and goats. In parts of Australia we can see the beginnings of new Saharas and the arid lands of the Middle East.

The agricultural balance sheet

> *I love a sunburnt country,*
> *A land of sweeping plains,*
> *Of ragged mountain ranges,*
> *Of droughts and flooding rains.*
> *I love her far horizons,*
> *I love her jewel sea,*
> *Her beauty and her terror—*
> *The wide brown land for me!*

So wrote Dorothea Mackellar, born into a country with one hundred years of white settlement and showing how some new Australians no longer thought of Europe as their home but had gained a sense of place in a southern world.

We are a wide brown land but our total area of good agricultural soils is much smaller than most people imagine. Limits on growing crops or pasturing livestock are set by the poverty of our soils, the rocky or saline nature of much of the inland and the variations caused by climate, particularly rainfall and heat. Cold is not quite the problem it is in the northern hemisphere.

Yet every country has its agricultural difficulties—cold, storms, flooding and the other vagaries of nature. In our 200 years of agriculture, for those farmers lucky enough to be in the fertile crescent from Cooktown southward into Tasmania and westward into south-eastern South Australia, with an outlier in Eyre Peninsula and then a big jump to south-western Australia, the land offered a rich living after the pioneering days were over. There are farmers in such areas who have never suffered major losses of crops or stock, generation

Water erosion. (M. Tanton)

after generation, though they have suffered from falling prices due to overseas influences.

A stroke of climatic good fortune in a country not richly endowed by nature is that we have a southern half with winter rains and a northern half with monsoonal rain. This allows a great variety of crops to be grown and some regions such as the Ord River have excellent soils and a future potential once suitable cash crops have been proved.

Mining the soil

It took many thousands of years before farmers came to realise that cropping land or pasturing livestock could not be treated as a mining operation. Good management was needed to plan for sustainable use. The massive clearing of the European forests had taken place about two thousand years before, and farmers coming to this new land regarded every tree as an enemy to be destroyed. They took it for granted that soils which could grow trees that dwarfed any in Europe must also be rich enough to grow magnificent crops. This did not always prove to be so. Forests depend on a recycling of nutrients; when the trees went, so did the soil fertility, transferred into the streams to a final destination in the sea.

Erosion, begun by unwise clearing, particularly on steep slopes, was worsened by too frequent cultivation. Livestock added to the problem by reducing soil cover over much of Australia's farmland.

Farmers soon realised the poverty of the soil. At Botany Bay it was found that the perverse Australian plant life bloomed on such infertility, a bonanza for botanists but a tragedy for farmers. So they turned to fertilisers, at first natural guano and later phosphate fertilisers imported from other countries and from some islands near Australia. A common joke in south-western Australia was that farmers grew their crops in the light soils by hydroponics but instead of the plants standing in water, they stood in sand, to which the farmer added the nutrients.

Today most farmers know there is an upper limit to the amount the land can produce, whether in terms of crops or by running livestock on the pastures. As a result of these hard-won lessons, and much research into suitable plant varieties for our climate, Australian farmers have earned an enviable reputation as among the most skilled in the world.

What is holding us back?

A federal system has inbuilt political difficulties. It is obvious that in certain areas the Australian government must be given overriding powers. Co-operation is increasing between state and federal authorities. In cases such as the Murray-Darling, which spans four states, there is need for concerted action and this has already begun. If any state will not take correct action, draconian measures should be used to ensure that one recalcitrant state does not do irreparable damage to what we still have.

Erosion

A commonwealth–state study showed that 200 years of white settlement had resulted in severe erosion problems.

Arid Australia

This is our heartland, covering 70 per cent of our area, the home of legend, the Outback. The fallacy that helped destroy the value of this region was the use of the word 'drought'. This term is used around the world as though it is something unusual, an act of God, a punishment. The punishment is actually due to the people who will not accept that in arid country there are dry times, medium times and wet times. In the dry times in parts of Australia no rain falls. In the wet times it would be impossible to grow crops, so steady is the downpour.

Yet these lands are marginal and nature took care of this over aeons of time. The water from the wet years was stored in the subsoil and drawn on by deep-rooted plants. Other plants used drought-resisting techniques such as dropping leaves, developing storage organs and the like. Drought dodgers shed seeds which remained viable until the rains came. Animals adjusted their lives or their numbers to the amount of food. The Aborigines moved around the country to take advantage of the less palatable but still nourishing food available even in the worst years. Occasionally the very young and the very old perished but the tribe survived.

Grazing by goats, rabbits and camels removed the plants at Eucla. Wind erosion then moved the sand dunes, overwhelming the settlement.

The desert lesson is still to be learned, particularly in Africa where so-called 'droughts' are devastating. Scientific advances have curbed the diseases which once kept populations low, so infant survival has increased. This has meant that people have to use their reserves of stock food, once eaten only during dry times, in the good periods. When the dry times come there is nothing left. Just like the populations of mice or grasshoppers, the human populations must also crash.

Sheep

These animals were the counters in the fortunes of many Australian squatters and it was not until the shocks of the 1890s that Australians began to realise that they could continue to ride on the sheep's back only if they stopped looking at how the sheep were doing, and looked instead at how the country was doing.

In western New South Wales, for example, the huge numbers of sheep that caused so much damage to native grasses and shrubs were slashed to a third. Added to the problems caused by these and other livestock there were also the new invaders such as rabbits, wild horses, wild goats and camels. All these contributed to the disaster to our soils. Myxomatosis saved us from complete calamity and better management has meant that sheep numbers are now usually kept at levels the country can stand.

The Aboriginal walkabout technique was used by cattle kings such a Sidney Kidman who moved his stock from station to station depending where the feed was. With modern transport farmers can do the same, shifting livestock to the districts which have had good rains. In many arid areas classified as having poor soils the wool clip is a far greater return than could ever be obtained from cropping.

Governments and graziers

The battle between governments and the holders of large properties is a continuing one, and some decisions have been more politically than environmentally wise. The attempt to cut down the size of holdings in many regions was a failure in both senses. A farmer with not enough land tends to overstock in order to earn a basic income and cannot practise the good management available to a larger landholder.

A new development has been the granting of large areas of land to Aboriginal groups. These people will need skilled guidance at first so that their properties will remain viable and produce enough for their needs.

Tourism is increasing in many parts of arid Australia and the setting aside of some land for national parks is the wisest use for it. Degraded land can then be restored, not only in terms of plants but also other wildlife. National parks in the western division of New South Wales are examples of how such overstocked country can return to something of its old beauty and interest in a decade.

Priorities

It is obvious that in this fragile environment the terms of land tenure must be adjusted so that the pastoral industry becomes more controlled and changes indicated by research results can be put into practice. Leasehold is much more flexible than freehold in this regard.

The problem of rabbits and other feral animals has already been mentioned, as has the problem of inedible shrubs replacing edible plants over huge areas of country. Preliminary research indicates that a new use of fire as a management tool will be needed to push back the advancing shrubs.

The high rainfall country

Although most of us think of grazing in terms of arid and semi-arid areas, nearly 60 per cent of our high rainfall regions are used for grazing on natural or near natural pastures. About 16 per cent of this huge area of some 13 million square kilometres needs erosion treatment due to past poor land use. The first warning of the damage grazing was causing in high rainfall areas came when scientists studied the alpine areas and found that the Man from Snowy River was not only going 'down the hillside at a racing pace' but that he and his fellows were taking the high country soil with them!

Last century the pattern of summer grazing the cattle and burning the bush to create new growth had become well established. Even in 1893 some scientists had warned the government that this was hazardous and was causing soil erosion in a fragile environment. The alarm spread from a few scientists to a larger number of people. Finally, with the use of the region as a water catchment, it was realised that soil erosion could not only threaten the grazing future but also the whole Snowy River Scheme.

In 1958 all land above 1370 metres was closed to grazing and at the same time, to make it even safer, all the high country was made a national park. Further safeguards came later and by this time Victoria and the Australian Capital Territory were also taking stock of what they had.

Gradually new national parks and more planning controls bought success. In 1986 all three governments agreed to plan co-operative management of this 1.2 million hectares of mountain country. This makes it one of the greatest of Australian national parks.

The Men from Snowy River are still fighting and a yearly gathering of dramatic groups of men and women on horseback produces sympathy for the graziers' cause. Those with expert knowledge do not share this feeling.

Having saved the high country, it is important we should now avoid mistakes in the tropical north which is facing the same kind of exploitation. Research is needed to find the best methods of pasture improvement, and make sure that graziers do not continue to mine the country.

Multiple use of catchments

For many years water catchment areas have been sacrosanct. Neither people nor livestock have been allowed in, to maintain the highest standards of purity. It is now realised that catchments are too valuable a resource to be used for one purpose only. It has been estimated in Victoria, for example, that the use of water catchments for growing timber would produce more jobs than the much vaunted aluminium works at Portland.

The same story is repeated around Australia. In the Sydney region early success has been obtained with plantings of red cedar. Cedar, the 'red gold' of the last century, was destroyed because of its value as timber. The industry also provided a cheap way to clear the forest for agriculture and often also obtain cheap land in the guise of a timber concession. A forestry myth was that red cedar could not be regenerated because of attacks by the cedar tip moth. I wondered how nature had managed to grow cedar at all, and tried planting a few in my garden, where they flourished.

What the myth really meant was that red cedar could not be planted as a monoculture as all such plantings are vulnerable to pest attack and the cedar tip moth caused heavy damage. Planted with other species as nature intended, the red cedar grows well. As the timber is so valuable, selective logging will in time produce high returns.

Another obvious use for water catchments is for recreation. This has always been important in Tasmania where the addition of foreign fish like rainbow and brown trout added to tourist attractions. Indeed the government advertises the new artificial Lake Pedder as a tourist attraction though in terms of beauty it does not equal the gentle elegance of the old lake with its shining white beach.

Tree farms

With the increasing loss of markets for traditional crops, farms should turn to this new resource. The value of eucalypts is well known but the Australian Centre for International Agricultural Research (ACIAR) is studying less well known trees and shrubs. This body has produced a book on multi-purpose use which gives information on 100 species of trees and shrubs with potential in this field.

Community forestry can use our sheoaks and wattles to add nitrogen to infertile soils and at the same time produce timber, firewood, charcoal, building poles and other material for many years from the same individual trees. We are aware of our international obligations and since 1983 all these potentially valuable species are being tested, not only in Australia but also in Thailand and Zimbabwe. What a paradox that our Australian Ark is producing, not only for us but for all humanity!

The arid, semi-arid and non-arid zones of Australia

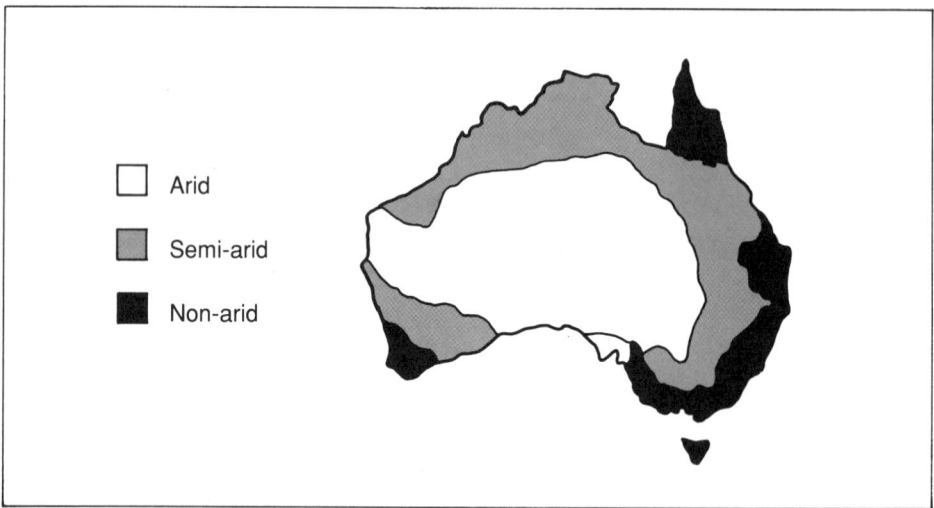

Source: Bolton, M. P., 'Parks and Reserves', in Messer, J., & Mosley, G., eds, *What Future For Australia's Arid Lands?*, Australian Conservation Foundation, 1983.

Land use policy for high rainfall regions

High rainfall regions are only a small percentage of the continent. As we have noted, Australia is the second driest continent in the world. Because of the intense heat of much of the inland, most Australians prefer to live near the coastline. This is one reason why we are the most urban of people.

Of the available land in this region, about 9 per cent is used for non-agricultural purposes—roads, factories and the like. It also contains most of our productive forests. Because of the rugged landscape much is unsuitable even for forestry but is ideal for national and wilderness parks, a valuable use of the land since it is close to the heavily populated urban areas.

The current land use of the rest is the most varied in Australia. The holdings are often small and are used for market gardens, orchards, vineyards, sugar cane and other purposes. In addition there has been some development of alternative lifestyle communities as well as hobby farms.

Conservation system

Governments should encourage changes in land use when present cultivation is unsuitable. A good example is the shift from dairying which has resulted because of competition from regions better suited to efficient production, such as New Zealand. Such land could be switched to tree farming which in the foreseeable future is assured of good markets both here and overseas. A wide range of suitable trees is available, including foreign and native pines, eucalypts, wattles and sheoaks.

Obviously tree farms can be heavily cropped only at intervals of thirty to fifty years but shorter term coppicing may provide some timber needs, including fuel, as is done in such countries as India. One eucalypt species, known there as the Mysore gum, but in Australia as the forest red gum, is cut every ten years in coppice fashion. This has been done for a hundred years with the original trees still thriving. Similarly thinnings of all kinds can be used for paper production.

Monocultures

A modern development has been the rise of factory style farming. In the worst examples large areas are cleared of all trees, and crops are grown on a grand scale often accompanied by irrigation and a massive use of machinery, pesticides and fertilisers.

Instead of animals grazing in paddocks or open enclosures, there has also been a development of indoor farming with hens, pigs, lambs and other animals never seeing the light of day. This raises serious ethical and aesthetic questions as well as other issues related to animal husbandry.

In the cotton-growing areas of New South Wales I saw families living in conditions that brought to mind the poor whites of southern United States. If the farm is turned into a factory, parents should not be surprised when their children prefer the factories of the cities to the factories of the country!

Ecological profit and loss

An item usually forgotten, both nationally and internationally, is how much energy is used to produce the amount of energy available in the food. In a peasant's world there is always an energy gain since the farmer depends on sunlight to grow crops and obtains nothing from outside the farm. In contrast, our much praised increase in productivity is often gained at the expense of the use of fossil fuels, which have a finite future.

We are often told how efficient European and American farmers are compared to those in Australia. For example, Australian wheat crops yield about one to 1–1.5 tonnes per hectare compared to the 7–8 tonnes produced in northern Europe. This seems to put our farmers in a poor light until one calculates the amount of energy used to produce that wheat.

In terms of fertilisers, pesticides, fuel for machinery and other factors, Australian farmers have an energy efficiency of 2.8, which represents a significant gain in energy by growing the wheat. Holland, the United Kingdom, Israel and the United States can offer an energy efficiency of only 0.5, a loss of one-half.

Australian farmers are storing the energy of sunlight while 'advanced' countries are wasting today's sun and also using the sunlight energy stored in fossil fuels such as oil and coal. Twice as much energy is used to grow the wheat compared to the food energy in the wheat.

However, when we calculate the energy used to get the food from the farmer's gate to your table the picture changes. Each unit of food energy available at the table requires five units of fuel energy to get it there. Eleven per cent of the energy gets the food to the farmer's gate, 38 per cent gets it from the farmer's gate to the store and 51 per cent gets it from the store to your table. What the world needs is more ecological profit and loss statements of this kind.

Feed the man meat!

A slogan good for farmers producing meat and for those selling it, but is it good for those eating it? As well as the health problems of too much fat meat in the diet, there is also the fact that the higher up the food chain an animal goes in search of food, the more wasteful it is in terms of energy losses.

Pesticide crisis gives us world's cleanest meat

. . . More than 1000 properties across five states are still in quarantine after being found to be contaminated with pesticides. Some may remain so for up to five years . . . Australia now has the most pesticide-free beef in the world with a background pesticide content of 0.24 per cent. This compares with 0.4 per cent in Europe and the United States . . .

August 1981: Horse and kangaroo meat found in beef exports to the US. Trade suspended, meat seized. Royal commission cites corruption in Federal Department of Primary Industry.

May 1987: Pesticide residues discovered in beef in the US and other countries. The US threatens to close market. Many farms in quarantine.

December 1987: Industry agrees to spend up to $26 million a year to clean up farms.

May 1988: More than 1000 properties still in quarantine. 614 120 animals have been tested. Australia has cleanest meat in the world.

(*Sydney Morning Herald*, 23 May 1988)

Plant variety rights

A hotly debated question in the last decade has been the question of plant variety rights. This means bringing breeding inventions under the same sort of legal controls as those that apply to mechanical inventions. In other words, having worked hard to produce a new variety of plant species which is distinct from the parents, with this distinction remaining stable over a number of plantings, the inventor may patent the work so that any other person wishing to use it must pay royalties.

At first glance this proposal seems perfectly fair but there is much to be said on the other side. There has been an instinctive reaction that anything provided by nature should not be patented. Then there are developing countries that have many species of plants but few of the experts to work on plant-breeding programmes. Having provided the raw material, they may find they have to pay royalties to grow a variety obtained from their own country. Until they have the experts such nations may regard it as a wise precaution to keep all botanists out of their country, which would be a sad day for agriculture.

Plant-breeding programmes are largely in the hands of big companies which are often also in the pesticide and fertiliser business. It could be desirable for these companies to produce varieties which, though high yielding, need pesticides and fertilisers for success.

Breeding for the public interest is not a way to make money so it will always be governments that search for varieties which need no costly extras to produce good results. There may be alternatives to such patents. Possibly governments could pay bounties for the inventor and then bring the new plant variety into the public domain. Alternatively the patent might last for only a short period. Our history shows that public expenditure on developing new plant varieties has paid dividends, so it might be best to continue in this way, although this would rob us of the opportunity to use the new plant forms developed in those countries which do have plant variety rights as part of their legal framework.

Pesticides

Dr Rachel Carson alerted the world to the dangers facing us from the new poisons being poured into the soil, the air and the waters of the earth with wild, technological enthusiasm. The dramatic results from poisons such as DDT, used during the war years against insect pests which had plagued humans by spreading diseases such as malaria, seemed to be ushering in a brave new world.

The first warnings of potential disaster came through Dr Carson's book. We learned that DDT and other persistent pesticides (in fact, they should be called biocides because they kill friend and foe alike) kept on killing along food chains. At the end were usually predators of some kind such as falcons, large fish and possibly humans themselves. So came the term 'silent spring' to describe a world where no birds sing.

The loss of our songbirds might have been accepted by farmers, except that as these and other natural controlling agents were being destroyed, so more and more pesticides were needed. When the birds stop singing in our gardens it will be a warning that humans are on the killing list of persistent poisons.

The world of the small also began the fightback. Insects breed much faster than the crops they devour, so many generations can be produced rapidly and among them will be individuals resistant to the new poisons. This resistance is passed on to their descendants. As the scientists produce new poisons, the insects respond with new varieties. We are on a dangerous rollercoaster to disaster.

Integrated pest management

In earlier times farmers had relied on three factors—a few poisons such as compounds of copper and sulphur, crop rotation to break the cycle of the pests and the acceptance of a certain amount of loss each year.

All these measures are being studied once again. In addition new biological controls are being discovered by research into the life history of pests. At certain points in their life a specific pesticide can be used. Today most of these are biodegradable—within hours or days of killing the pest, they change into harmless materials.

Cactoblastis is mentioned elsewhere in this book. Ladybirds as destroyers of aphis have a long history. The green vegetable bug was once a serious pest in the market gardens. A wasp was introduced whose female laid her eggs in the eggs of the bug; the bug eggs were then eaten by the wasp larvae. Dozens of such biological methods are being used today. Research into such animal friends is slow and costly but once achieved, works forever with little cost to the farmer.

There can be mistakes. In the 1930s farmers insisted that cane toads be brought into Queensland to control a beetle that was a pest on sugar cane. Some scientists warned that more study was needed but the toad was introduced nevertheless. It served no purpose in controlling the sugar cane beetle but it did eat practically everything else, including honeybees and a variety of small creatures which add interest and variety to our wildlife as well as eating pests themselves. Today the cane toad is the most common vertebrate in coastal Queensland. It is a nuisance in urban streets, kills domestic pets, and is spreading south into New South Wales and west into the Northern Territory.

Resistant plants and animals

This control method has been used for centuries. The disease phylloxera almost wiped out the vineyards of the world. It was found that some low yield grape varieties were resistant, and by grafting high yielding vines on to the low yield root stock the problem was solved.

Zebu cattle were brought into northern Australia as they thrived in the hot conditions and were resistant to the effect of cattle ticks.

Other methods

Much ingenuity has been used to attack pests. Always there is the need to know the enemy. First it is necessary to identify the species or variety. This is the job of taxonomists and is why the seemingly esoteric study of taxonomy has great value.

Knowing your enemy also means studying its life history. This includes any controlling factors in the home country. If all else fails it may be possible to change cropping methods to reduce damage. Perhaps the harvest can be taken when the pest is at its lowest ebb. In the case of codling moth which attacks apples the tree trunks can be scraped to remove pupation places. The moth larvae then have difficulty in finding a suitable spot to rest before emergence.

Integrated pest management allows a farmer to use a variety of control methods which also cut costs.

Poison waste disposal

New poisons are continually coming on to the market and it is important that they be used exactly according to instructions. Labelling must be clear and simple. The disposal of unused material is an important factor. Too often it is tipped into drains or creeks, where the killing continues downstream.

Poison waste disposal is an area where governments must move rapidly since the potential for disaster is always present. American scientists working in this field say they would prefer taking the chance of living near a nuclear reactor than the same distance from a dump accepting poisonous wastes!

An interesting point of view has been put forward by a prominent leader of Australian farmers. He points out that, having no nuclear facilities, we could sell our products around the world with the reassuring label that they come from a nuclear free continent.

Above all, governments must work hard at finding solutions so that pesticides will be phased out of farming. This kind of research cannot be left to industry since it is only commonsense that a maker of pesticides is not in the business of making the factory bankrupt.

Farm products are already being advertised as coming from land where no poisons or other chemicals, growth hormones, etc., are used. What a price advantage we might gain if we guaranteed all our exported food was 'pure'.

Fertilisers

In the infertile soils of Australia these artificial aids have been essential to success. Today, because of the expense, most farmers are well aware of striking the right balance between too much fertiliser and too little. In addition, research has shown the importance of many trace elements without which stock will not thrive or may die.

The farm bank

The seven million hectares being used for other purposes needs careful examination by governments. Valleys disappear to create water storages and land is set aside for national parks and recreation areas and forestry. In addition, thought must be given to the much delayed restitution to the Aborigines of the land taken during the early days of settlement. Government agencies such as the army, keen to remain near the larger cities, look for land nearby, a wasteful use of a valuable asset.

The urban wave that devours farmland needs curbing. The popular belief that cities cannot be stopped from growing is a myth encouraged by powerful lobby groups and politicians who want a quiet life. For those who own the land an increase in population means an increase in wealth. It is a case of private affluence defeating public good. Decision makers who have no desire to leave their comfortable homes produce reports showing how wise it is to push more people into a smaller space, or let those who cling to the Australian ideal of a breathing space around each house, spread into the nearby countryside.

Despite this philosophy, new cities have grown. Canberra is an example of a city created by government edict, and the belief that it is a dull place in which to live is not shared by those lucky enough to live there. Mining towns have appeared like mushrooms across Australia while cities such as Townsville indicate how an injection of government money, in this case by shifting the army nearer the potential front line and building a new university, can cause a steady growth.

People go where the jobs are! The myth that people like to live in giant cities has been punctured many times. Many years ago that dean of city planners, Lewis Mumford, showed that a city of a million people can offer all the urban attractions any sophisticated person needs. A recent poll showed that 40 per cent of people under thirty would like to move out of Sydney if suitable jobs were available.

Conservation objectives

The conservation objectives for ideal agricultural production include:

1. The study of soils, landscape and climate to ensure that the right kinds of crops are matched to the right kinds of country, both nationally and internationally.
2. A close monitoring on soil fertility, particularly with regard to correct use of fertilisers and pesticides. Alternative lifestylers should be encouraged to take rundown country and restore productivity, particularly by recycling waste products instead of using artificially produced fertilisers, a practice once common around the world and still to be found in many countries.
3. A much closer partnership between state and federal governments.
4. A national tree policy not only to restore old forests but to encourage such plantings as will achieve a tree cover that may resemble pre-Aboriginal times. In many cases it is not so much a replanting that is needed, but the temporary fencing of cleared areas to allow the original bushland to regrow. This has met with great success in some parts of Australia, the most notable examples being Broken Hill and Kalgoorlie.
5. Prime agricultural land must never be destroyed unless there is no prudent or feasible alternative. The attempted army takeover of land in the Orange–Bathurst and Cobar regions was a good example of unwise planning.
6. Every country should be able, if the worst comes to the worst, to feed its own population. A nation can survive as long as it produces enough food for its needs. Food prices are now low, but it must be remembered that, just as the oil producers formed a cartel and forced up prices, the producers of food may at some time in the future form a cartel and become dominant, instead of being the most helpless when it comes to marketing.

Feral animals and plants

Bringing a new plant or animal into a country is a two-edged sword. Food plants and stock animals are often able to leave behind the diseases and pests which plagued them in their home countries but unfortunately, during the last century, a number of plants and animals were introduced with tragic results.

Conservationists dream of a sorcerer who could wave a magic wand and banish all feral animals from Australia. The same applies to 'feral' plants, those unwanted species that have gone wild, including pests such as prickly pear, lantana and privet.

Where animals are concerned, emotion sometimes clouds reason. Killing animals like rabbits, horses, donkeys or camels is not pleasant. Yet these creatures are destroying our environment at an alarming rate. The ecological damage done by the rabbit is well known but the others are also dangerous. Wild pigs, for example, were the cause of the near extinction of the Lord Howe Island woodhen. In the Northern Territory there are around a million feral horses, pigs, water buffalo and donkeys and to this figure we must add camels, rabbits, foxes, cats and wild dogs. The losses these animals cause are astronomic.

All conservationists agree that these feral animals have to be destroyed but ask for this killing to be as painless as possible. Where practical, the carcases should be used for some economic purpose; however even leaving them on the ground is not a waste since the material in their bodies is recycled into the soil.

Weeds

Small herbs have grace,
Great weeds do grow apace.

In *Richard III* Shakespeare used an example from natural history to drive home a message. Yet what is a weed but a plant in the wrong place? In a natural community there can be no weeds as all the plants are part of the web of life of a particular region.

Some idea of the number of introduced plants in Australia is given by the fact that 10 per cent of all our plants are exotics, and the number increases steadily. A few native species, with the helping hand of humans and domestic stock, have been given an advantage over their fellows and may be called 'weeds'. Native bracken spreads after clearing. In inland areas the vast plains of saltbush and blue bush have been replaced partly by native grasses and other native shrubs, when the earlier plant cover was grazed out of existence by sheep.

A few aliens can invade an undisturbed natural community. Boneseed or bitou bush is one. Fortunately most of our bushland resists the invaders unless humans aid them. Many of our weeds are plants brought here to adorn a landscape many settlers found dull by comparison with Europe. Think of that grand old man of botany, Baron von Mueller, walking eagerly across Victorian valleys pushing his blackberry canes into the soil, in an imitation of warlike Admiral Collingwood who carried acorns in his pockets, so that in every suitable English woodland he could push in the seeds that would one day grow into stately oaks to provide the timber for the hulls of an English navy!

Then there is Paterson's curse, a garden plant that escaped into paddocks of southern Australia. Another reason for the success of some native, and most of the invading plants, is that they thrive on disturbed soil. The firestick farming of the Aborigines helped to increase particular native species and these, with the aliens, became the weeds.

Another factor is also important. When it leaves its home a plant also leaves those natural controls which kept it in check—competition either from plant fellows or from predators, of which insects were the most common.

Most exotics find it difficult to penetrate undisturbed bushland, but such an area is becoming a rare commodity. Rabbits, hares, sheep, goats, camels, horses, donkeys, pigs, rats, mice—all have played their part in opening pathways for invaders.

Even more striking is the effect of fertilisers in assisting newcomers. Botanist Professor John Turner commented that even a piece of orange peel or a cigarette butt can increase soil fertility in heathlands and cause weeds to enter. Added to this is wind drift of fertiliser spread from planes that reaches bushland nearby.

The prickly pear is a marvellous example of a plant invader. For a time it took vast areas of farmland out of production. Cacti were carried to Australia with the First Fleet but the species which did the damage was brought in later as an ornamental plant. It found conditions in northern New South Wales and southern and central Queensland ideal. At the height of infestation some 26 million hectares of agricultural land disappeared under the green sea of cacti.

The prickly pear was attacked in the traditional ways of slash and poison but scientists knew that there was only one economic solution. They went to the home of the prickly pear to search for predators. A number of insect devourers of the cactus were found and tested to make sure they would not become pests on crop plants or native plant life. Several proved useful in different parts of the prickly pear range but a moth named *Cactoblastis* was the real conqueror. The caterpillars, the larval stage of the moth, viewed the expanse of prickly pear like children staring at sweets and soon ate away the green fields of cacti.

As the old farms emerged agriculture returned. When the major pear infestations were destroyed there were always enough moths produced in small patches of pear to bring under control any large outbreak which developed in later years. Since those happy days some problems have arisen with new species of pear but these are not a disaster compared to the earlier outbreaks. Farmers were so appreciative of the work of the moth that they named a shire hall in its honour!

European rabbit

There is a persistent belief that rabbits were brought to Australia from England because the settlers were not aware of their pest status. Any English farmer could have told them the truth. Rabbits came with the First Fleet but were domesticated varieties used for food. Those that escaped did not thrive in the coastal settlements. It was Thomas Austin who in 1859 had wild rabbits trapped on an English farm and was able to acclimatise them in this new land for some sporting shooting.

They became a grey horde which swept over two-thirds of the continent. Austin was the sorcerer's apprentice who began the flood but where was the sorcerer who could stop it? It is a fascinating story and I recommend all interested to read Eric Rolls' book, *They All Ran Wild*. For those who think that conservation groups, the 'greenies', should leave all our problems in the capable hands of public servants, it is a salutary lesson.

In the 1920s a Brazilian scientist suggested a solution to the rabbit problem by using a disease which destroyed South American rabbits. Early efforts were lackadaisical and it was only by the determination of Dr Jean Macnamara that Australia was dragged into the modern world of pest control. She was an expert in viral disease. When the federal government decided that lack of early success meant that further trials should be dropped, they had not taken into account Dame Jean. She used Victorian politicians as her battering ram. If the federal government would not move then Victoria would go ahead without them.

Above: *Prickly pear photographed in April 1928, prior to attack by the cactoblastis insect. (Department of Lands, Queensland)*
Below: *Eighteen months later, in October 1929. (Department of Lands, Queensland)*

The soil has taken a terrific beating as the native vegetation has been replaced by annual crops and by pastures for grazing ...

A further insidious soil degradation is only now becoming slowly visible, sometimes from salinity, sometimes from acidification. Salinity is usually caused by poorly considered irrigation or by excessive clearing of timber. Acidification is the result of decades of using fertiliser. Huge areas of our prime cropping and grazing country have now become too acid for healthy growth ...

One of the most fundamental agricultural problems, and one that has been given almost no heed in the past, is loss of nutrients as agricultural products are exported to distant cities, both in Australia and abroad ...

It is a one-way street from the wheatlands of Wagga to the sewer outfalls of the world.

Nor must we ignore the huge quantities of nutrients which continue to be lost under the impact of wind and water erosion. Under existing practices each tonne of wheat means a loss on average of 10 tonnes of soil ...

In 1986–87, the wheat crop of 16.1 million tonnes removed 322 000 tonnes of nitrogen and 48 000 tonnes of phosphorus from the soil. This same wheat crop also removed 64 000 tonnes of potassium and 32 000 tonnes of sulphur.

Although nitrogen is replaced in the soil by rhizobia bacteria on the roots of subterranean and other clovers stimulated to grow by superphosphate, the traditional 'sub and super' combination is what has led to acidification in the fertile and well-watered areas.

It takes tonnes of lime to counteract the effects of hundredweights of super-phosphates, and that has to be applied before the acidification reaches the subsoil ...

We have a culture that does not know how to respect and look after its land ... radical changes must be instituted to stabilise population levels, encourage decentralisation, foster self-reliance, increase recycling, and promote the stewardship of the land.

These are the issues of the '90s.

(Dr Chris Watson, CSIRO Soil Scientist, in the *Bulletin*, 31 January–7 February 1989)

Federal research continued until there came a day which staggered the scientists. I remember my brother, who worked in the CSIRO, telling me how many of the research team had little faith in myxomatosis and were issuing a report on how the test had failed when suddenly stories came of rabbits dying in millions along the Murray River. In some places up to 90 per cent of the population died.

The disease quickly spread, by natural means and by the deliberate introduction of infected rabbits, over the whole of Australia.

The story is not quite as simple as I have told it. Like all viral diseases, the effectiveness of myxomatosis varies and other methods must be used to keep rabbits in check. Yet Dame Jean Macnamara was the real sorcerer who turned back the grey flood and allowed farmers a breathing space to get their defences in good order so that years later they were able to handle the rabbit problem when it returned.

A side effect of the rabbit numbers was an abundant food supply for feral cats. These domestic pets went feral in the early days of settlement. When rabbits began to spread in plague proportions many farmers thought of cats as possible control agents. A brisk trade began in selling them to farmers but it is not a simple matter to establish a domestic cat in the bush. It is naive to believe that if only humans would stop dumping kittens in the bush the feral cat problem would disappear. Kittens dumped in the bush die of starvation with the lucky ones being killed by predators that include feral cats. The bush is completely populated with cats and for another individual to gain a place is rare. Even today feral cats depend heavily on wild rabbits for their food.

An even greater disaster for our native animals came from the release of European foxes to provide sport for the gentry. The fox proved a devastating predator on Australian wildlife. I have even found them digging into the ground for native land snails. Their ability to eat small creatures enables them to thrive when other predators starve.

Other grazing competitors to our wildlife include feral goats, donkeys, horses, camels and house mice which often occur in vast plagues. All these animals have created major problems in parts of Australia.

Feral pigs

Captain Cook carried a store of live pigs which not only served as food but could also be used for testing new plant products. If the pigs survived a new fruit or seed it would be safe for humans! During his enforced stay on the Endeavour River some of the pigs escaped and probably became feral.

Similarly, in the early days of white settlement escapees from government stocks became the nucleus of new colonies in better watered areas. With the capacity of becoming mature

Camels run wild in arid Australia.

in seven months, and a sow being able to give birth to ten piglets every three months when conditions are good, the problems for farmers became acute.

The major damage is to the plant life of a region but at times pigs will turn from vegetable food to young animals. Dr Jack Giles of the New South Wales National Parks and Wildlife Service found that in one area of northern New South Wales pigs were killing 92 per cent of the lambs in one season. Much more alarming is their potential ability to spread diseases which can affect both livestock and humans. Should foot-and-mouth disease enter the country from the north pigs could carry it throughout Australia. Dr David Sexton, once chairman of the Victorian Vermin and Noxious Weeds Destruction Board, wrote:

> Although the likely role of feral pigs as a significant reservoir for foot and mouth disease and probably rabies is well known, there seems a disturbing degree of complacency about this threat. What is perhaps less well known is that feral pigs act as host for diseases, such as brucellosis, leptospirosis and Australian encephalitis in some parts of the country, not to mention swine fever . . .

Numbers are hard to estimate but it is suggested there may be three million pigs in New South Wales and four million in Queensland, with the damage to the pastoral industry placed at about 100 million dollars.

The success of the Lord Howe Island woodhen programme was due in great part to eliminating feral pigs from the island. On the mainland this is impossible. There is a market for feral pig meat in Germany but the only method of control will be a steady war of attrition. The best that can be achieved will be to keep the problem in manageable proportions, the kind of solution which applies to most of the pests in this country. We can never hope to exterminate but at least we can reach an acceptable level of damage compared to the cost of control.

Native animals as pests

Since the earliest days of white settlement, farmers with their domesticated animals faced two problems. First and most obvious were predators such as Tasmanian tigers, devils, quolls, wedge-tailed eagles, crows and dingoes which attacked stock such as poultry and sheep. Cattle, being too large for native predators, were reasonably safe except from Aboriginal spearing.

A relentless war was waged against such attackers, including human, as Aborigines were swept into oblivion in most settled areas. The Tasmanian tiger was hunted to extinction although the other predators, being smaller and more numerous, managed to survive. Farmers resented any other grazing animal taking any of the native grasses and other plants or feeding on planted crops and pastures. Most kinds of kangaroos, emus, Cape Barren geese and other plant eaters were harried with gun and poison.

All the research so far carried out on so-called pests has indicated that they are not a serious difficulty for farmers. Many hunters are, in fact, of assistance to graziers since their main diet consists of more important pests such as rabbits. Most 'pests' are at worst only a nuisance and the time spent pursuing them should be spent on developing better farming

Grey kangaroos at Sperm Whale Head National Park. Two kinds of grey kangaroos and the red kangaroo and the euro are the four largest kangaroos still being killed as pests. (See also pp. 98--104)

Water buffaloes in Kakadu National Park. These introduced animals soon went wild and over the years devastated the wetlands of this national park. Over recent years the animals are being removed and much of the old beauty is returning. (See also p. 90)

Wedge-tailed eagle. Until recently all birds of prey were regarded as enemies and giants like this eagle were hunted and poisoned as pests. Today research has shown that so called wildlife 'pests' are usually minor problems at particular times of the year and often, as in the case of the wedge-tail, useful as killers of rabbits. (See also p. 96)

The emus won the guerilla war!

methods. Newborn lambs protected by shelter belts can survive. It has been found that all lambs less than two days old die when exposed to cold or wind. Often this kind of loss is blamed on crows, eagles or foxes when it is basically a climatic problem. The predators arrive after the lamb is dead. They are scavengers rather than killers.

Two case histories follow: the emu because it shows how ecological commonsense can solve problems, and the kangaroo because this problem is still hotly argued.

The emu war

The emu is a good example of how research may solve a minor agricultural problem in a way which satisfies any reasonable person.

This tallest of all our birds, though honoured on the Australian coat-of-arms, had little honour among farmers who believed that it competed with stock for available food. In cropping areas it caused damage not only by eating heads of wheat but by trampling the crops so they could not be harvested easily. When panicked, birds damaged fences in their headlong flight. They also damaged themselves!

In the days of the prickly pear problem the emus were blamed for spreading the seeds of the prickly pear. Emus eat fruit in season and since they can travel great distances they were undoubtedly one of the many creatures which helped spread the pear. State governments offered bounties for killing emus and collecting eggs. Hundreds of thousands of birds and eggs were destroyed before other methods were discovered to counter the prickly pear menace. So the emu, since the earliest days of farming, was harried, first as a source of fresh meat then as a destroyer of crops.

An extraordinary story in Australian conservation comes from Western Australia where harassed farmers appealed to the state government for aid. It decided to call in the Australian Army. The federal Minister for Defence was Sir George Pearce, a West Australian politician dependent on farm votes who expected to gain political advantage from the agreement. So began the Emu War.

A detachment of the Royal Australian Artillery was sent to do battle with the enemy which was forming into groups numbering many hundreds of individuals. The soldiers were armed with two Lewis guns and 10 000 rounds of ammunition so it was thought this would be a fight easily won.

Emus usually avoid visual contact with each other, being solitary birds, and it was only hunger and thirst which had created the mobs. The opening fire sent the emus in all directions and they reformed into small groups in the classic guerilla pattern. The Lewis guns jammed at a critical moment and the main body escaped.

The war continued with little result. Then an emu ally entered the fight. Sir George found that local conservationists began to harry him while protests poured in both from home and abroad. Even more important, the government was becoming a laughing stock and ridicule caused the army to be withdrawn.

Other methods of control were sought and research undertaken to see where the problems really lay. It was found that many emus, after nesting in the native pastures of the north in winter, wandered southward on a migration that often covered hundreds of kilometres. This brought them into the paddocks of the farmers who were cropping marginal lands and who rarely had the money to build fences. In summer the survivors of the southern trek moved north once more to join that part of the population which had stayed put and they bred once more to send new contingents south at the end of the next winter.

The building of vermin fences to stop the rabbits also stopped some emus, and by an extension of such barriers they blocked the birds' move south. So the emu ceased to be a significant pest. Those living on the southern side of the fence were much more easily handled by the farmers themselves. The result is that the farmer was saved time and money. The small number of emus left added interest to the countryside. A new direction has been the farming of emus by an Aboriginal group at Wiluna, Western Australia. Both feathers and high grade leather are produced.

The kangaroo problem

Few conservation questions in Australia have caused so much argument, so much anger, so much confusion as that of the kangaroo killings. Here is our national emblem, recognised the world over as the symbol of this country and enshrined in our national airline, being killed in millions each year. What is a kangaroo? Forty-eight kinds of marsupials are included under this broad name, including bettongs, potoroos, wallabies and kangaroos. The following list indicates the species which are destroyed as 'pests' or taken for sport, at least in some parts of their range:

red kangaroo *(Macropus rufus)*
eastern grey kangaroo *(Macropus giganteus)*
western grey kangaroo *(Macropus fuliginosus)*
euro or wallaroo *(Macropus robustus)*
whiptail wallaby *(Macropus parryi)*
red-necked wallaby *(Macropus rufogriseus)*
agile wallaby *(Macropus agilis)*
black-striped wallaby *(Macropus dorsalis)**
tammar wallaby *(Macropus eugenii)**
swamp wallaby *(Wallabia bicolor)*
red-bellied pademelon *(Thylogale billardierii)**

*These have a restricted range but are sometimes a pest in a few places.

In general it is the six largest species that provide the major harvest for kangaroo shooters. Are any of these kangaroos in danger of extinction? The following list of extinct and endangered kangaroos is that accepted by most conservation groups:

boodie or burrowing bettong *(Bettongia lesueur)*—vulnerable
woylie or brushtailed bettong *(Bettongia penicillata)*—endangered
desert rat-kangaroo *(Caloprymnus campestris)*—presumed extinct
longfooted potoroo *(Potorous longipes)*—endangered
broad-faced potoroo *(Potorous platyops)*—extinct
central hare-wallaby *(Lagorchestes asomatus)*—extinct
rufous hare-wallaby *(Lagorchestes hirsutus)*—vulnerable
eastern hare-wallaby *(Lagorchestes leporides)*—extinct
banded hare-wallaby *(Lagostrophus fasciatus)*—vulnerable
bridled nailtail wallaby *(Onychogalea frenata)*—endangered
crescent nailtail wallaby *(Onychogalea lunata)*—extinct
proserpine rock-wallaby *(Petrogale persephone)*—vulnerable

From this list it becomes clear that none of the larger kangaroos is at risk at present; it is the smaller species which have suffered most. Often this was due not to direct action by humans, but an indirect result of land clearing and grazing. Sheep not only competed for food but also removed the low shrub cover essential for the smaller marsupials.

People do not live by bread alone; neither do wild animals. Sheltering holes in trees are necessary as homes for many of our tree-living mammals, birds and reptiles. Similarly a covering of shrubs is necessary for those animals which spend their lives on the ground to provide hiding places from enemies and shelter from heat, cold or drying out.

Estimates of the red, eastern grey and western grey kangaroos total around twenty million. Their numbers fluctuate wildly depending on good and bad seasons. As kangaroo harvesting has been in the order of several million annually and this has been continuing for several decades it is obvious the population is huge in most years.

In the 1960s very heavy cropping combined with dry seasons brought kangaroo numbers so low as to arouse conservation alarm. The bland assurances of various government wildlife authorities were not accepted and, as a result of lobbying, a ban on the export of kangaroo products forced all states to agree to bringing in new regulations which have since provided better management in this industry.

Have the populations of kangaroos changed since white settlement? Conventional wisdom is that the numbers have gone up. An obvious cause is the decimation of the Aboriginal population whose hunting must have resulted in the killing of hundreds of thousands of kangaroos each year.

The killing of dingoes by farmers removed one of the major predators of kangaroos. There is recent evidence from Queensland, where a dingo fence separates very similar country, that in the region where kangaroos and dingoes are not molested the kangaroo numbers are much less than on the other side of the fence where dingoes have almost been eliminated.

Food preferences

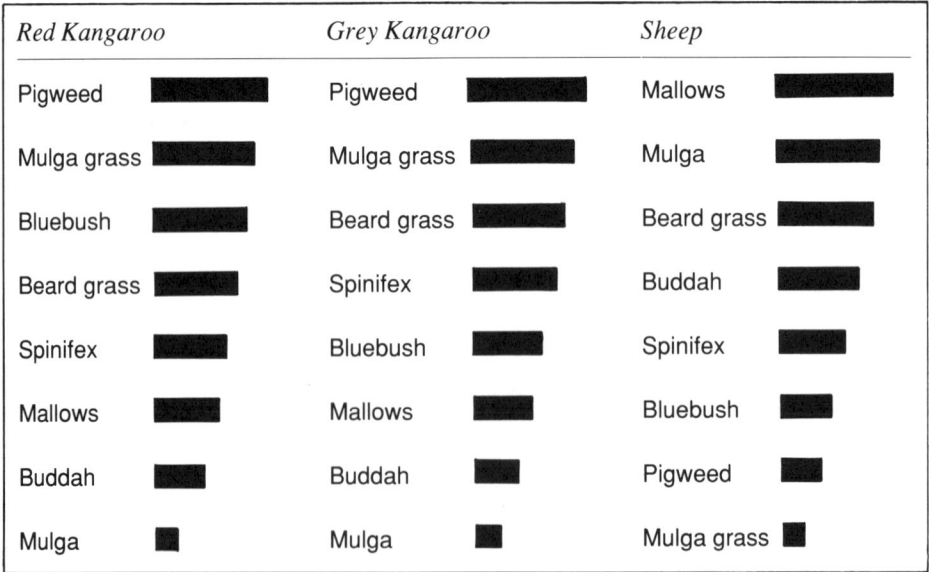

Red Kangaroo	Grey Kangaroo	Sheep
Pigweed	Pigweed	Mallows
Mulga grass	Mulga grass	Mulga
Bluebush	Beard grass	Beard grass
Beard grass	Spinifex	Buddah
Spinifex	Bluebush	Spinifex
Mallows	Mallows	Bluebush
Buddah	Buddah	Pigweed
Mulga	Mulga	Mulga grass

Source: Biological Science: The Web of Life.

The extra watering points needed for domestic stock have improved conditions for the kangaroos and helped in their survival, though it must always be remembered that food as well as water is necessary for any animal population. The selective grazing of sheep removed plants such as saltbush and other shrubs not favoured by kangaroos and allowed a greater growth of grasses which are favoured by kangaroos. A striking example was the case of the euros in the Pilbara district of Western Australia where grazing by sheep allowed the native spinifex, a type of tussock grass, to increase. On this food the euro survived—but not the sheep! Closer to the coasts the gradual fencing of paddocks and the more intensive killing by farmers and sporting shooters almost eliminated all kangaroos in the more settled areas.

Yet there are some conservationists who, after a study of explorers' field notes and accounts by settlers, claim there is evidence, at least in some areas, that the large kangaroos were as abundant then as they are now. So the argument that it was European settlement that created the kangaroo problem may not be valid.

Why are kangaroos killed?

Much of the material in this section is taken from a paper on kangaroo management produced by the Australasian marsupial specialists group of the International Union for Conservation of Nature and Natural Resources (IUCN).

Governments usually state that the only justification for kangaroo shooting is that they are agricultural pests so their numbers must be reduced, and that it would be a waste of a valuable resource to allow their bodies to be left lying as is done in Victoria at present. The carcases have value for pet meat and human consumption, for fur and leather.

Some conservation groups put forward two contrary views:

1. The existence of a kangaroo industry produces a strong lobby group to encourage the harvesting of kangaroos on a regular annual basis even in years where the animals are not in pest proportions.
2. There is also a strong body of opinion that there is no reliable scientific evidence to indicate that in the two-thirds of Australia which houses the pastoral industry depending on native grasses, the total removal of all kangaroos would increase the numbers of domestic stock the country could carry.

Kangaroos are not the only wildlife which graze native plants; grasshoppers, termites and other insects play their part. Indeed the farmers may be wasting their energies on the wrong enemies. The situation is complex but there is a great deal of evidence to show that only in drought years do kangaroos and sheep compete for the available plant food. In drought it might be wiser for farmers to reduce their stock numbers for their own long-term gain.

Numbers of kangaroos

The IUCN statement on this matter was:

According to recent aerial surveys and other estimates the Australian population of red, western grey and eastern grey kangaroos in 1981 was conservatively estimated at 19 million. Thus a commercial harvesting limit of two–three million represents approximately 15 per cent of this population. Under the best seasonal conditions kangaroos can increase at 30–40 per cent per annum so the past commercial harvest appears to be within safe limits. Nevertheless continuous population monitoring is essential to take into account the effects of habitat changes, drought, pest control by farmers and shooting by sportsmen.

Some conservation groups however doubt whether the official figures of kangaroo harvesting are accurate in some states. Anecdotal evidence indicates that some official records may be inadequate.

Is kangaroo hunting cruel?

Full-time professional shooters must be accurate if they are to remain in the industry. The community tolerates battery hens and other new forms of factory farming relating to pigs, calves and sheep, the deaths which take place in transport of farm animals (recently estimated by the RSPCA to be at least 1 per cent of sheep during transport) and other aspects of animal husbandry on farms. There can be no question that kangaroo culling is the least cruel of current farming practices. What should be done?

The IUCN report made a number of recommendations for improvement to kangaroo management:

1. Find out the habitat requirements of the species which are being culled and note any changes which may affect the numbers.
2. Areas should be studied to find where kangaroos can be allowed to increase, must be kept static or need reducing. Such decisions will depend on whether the area is a grain-producing region or is one where stock are feeding on native pastures.
3. Select areas in which shooting will be allowed.
4. Check both the size of the populations and how many are being harvested. Results should be made public.
5. The harvesting rates should be such that the numbers taken can be sustained from year to year.
6. A minimum density should be set for every region. When numbers fall below this level harvesting should stop.
7. The killing methods should be such that cruelty is kept at the lowest possible level. Inevitably when any animal is killed there must be some pain but by the use of professional hunters this will be kept to an acceptable level.
8. Greater use of fencing would keep kangaroos from agricultural zones though care will be needed to study what effects such fences might have on the populations of kangaroos in the area.

There is no doubt that these are good ground rules and once the general public is convinced that governments are carrying out these recommendations in practice as well as in theory the clamour for kangaroo killing to end will be greatly reduced.

Above all there must be an adequate development of suitable national parks or similar reserves. In such places the public will be able to watch these graceful creatures in their natural state. Even with correct management, in other areas the kangaroos will suffer from

How kangaroos are counted

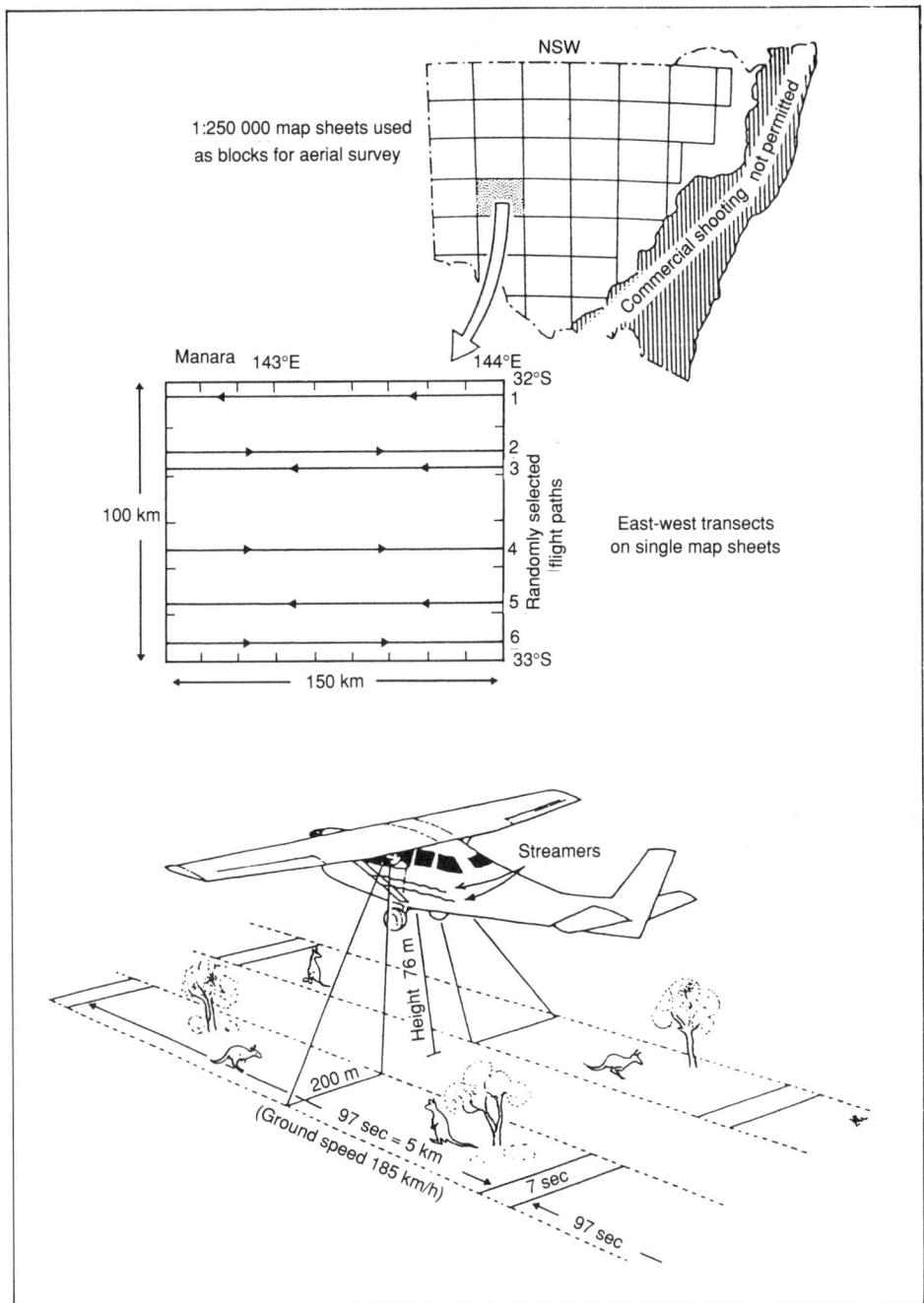

Source: Department of Arts, Heritage and Environment.

visual extinction since regular hunting will keep them well away from main roads. The sound of a car will be a signal for the animals to move into hiding.

It is this behaviour pattern that has led to the idea that kangaroos are rare even when accurate counts show they are plentiful. Most travellers are aware of the change in behaviour seen in kangaroos protected by national parks. It is analagous to the tameness of ducks in city parks compared to the same individuals once they return to the wild.

In summary, there is no chance of the larger kangaroos becoming extinct though the same confidence cannot be held with regard to the smaller species. Given an adequate number of national parks spread evenly across the countryside every person will be able to study all our larger kangaroos in comfort.

Should any wild creatures be harvested?

This is a new departure in the conservation movement. For vegetarians the answer is simple. Most accept the Buddhist concept of reverence for life and treat all animals as sacred. Yet to protect plant crops many pests must be destroyed. It is difficult to accept a philosophical concept which has a dividing line among animals: invertebrates such as insects and snails may be destroyed while backboned animals such as frogs, fish, reptiles, birds and mammals are protected. And what of disease bacteria? Are they to be revered?

So for most Australians this ethical position is confused. We object to the killing of whales, dolphins and seals, all creatures with an appeal to humans. This often extends to the demand for banning the capture of dolphins for display in aquatic pools, yet the caging of equally intelligent wild cockatoos is condoned. On the mainland brushtail possums are protected but in Tasmania they are hunted for their skins. Seabirds in general are protected except for the short-tailed shearwater which is harvested as the Tasmanian muttonbird.

Many other examples could be given but there is obviously a slow shift away from the exploitation of wild animals of any kind, although in the future many may be farmed under captive conditions. This is already being done with emus. It has also been suggested that it would be possible to obtain a greater return from native pastures by harvesting kangaroos and wallabies rather than confining a farmer's attention to sheep or cattle.

Even with the best management there will still be the need for culling certain animals in particular regions when a combination of circumstances produces pest proportions. Illogically there is no outcry when plagues of native rats or house mice are killed. Yet feelings of compassion are important to most humans. The simplistic scientific attitude that animal deaths are part of nature's pattern is not acceptable to most Australians, who demand that efforts be made to save stranded whales, oiled seabirds, injured koalas and the like.

Quarantine

An efficient service and enabling laws have helped keep unwanted animals and plants out of Australia during the last four decades. Gradually loopholes are being closed, although even the most stringent of precautions can be circumvented by humans who go to extraordinary lengths to conceal illegal materials.

The use of water ballast for ore ships returning to Australia increases the chance of overseas marine plants and animals being introduced to Australia. An even more alarming situation is provided by fishermen, particularly from Indonesia and Taiwan. While tourist boats are under strict regulations regarding the dumping of food rubbish overboard when in Australian waters, it is difficult to police smaller boats. Indonesians often camp illegally on offshore islands. At times they have been discovered in camps with domestic fowls. This provides an opportunity not only for the entry of poultry diseases but also viral diseases of humans. The Australian government should cancel all letters of understanding which allow such fishermen to land on any of our mainland shores or islands.

Forestry

What did we have?

In 1788 major forests covered about 10 per cent of Australia with woodlands covering about 23 per cent more. In the next 200 years we removed two-thirds of these trees, often through necessity to grow crops and pasture cattle, sometimes through a sheer bloody-minded hatred of the bush. It will be a long, slow and costly road to repair the mistakes of the past.

Forests of Australia

Australian indigenous forests cover 41 million hectares (5.4 per cent of total land area) of which 30.2 million hectares are publicly owned and 11.0 million hectares are in private ownership. Of the public forest areas 4.9 million hectares are in national parks. In addition there are 842 900 hectares of forest plantations, 802 500 hectares of which are coniferous (exotic and indigenous softwoods). The current planting rate for these plantations is 33 000 hectares per year. Woodlands cover an additional 64 million hectares of Australia (8.3 per cent of total land area).

(National Forest Strategy)

Where have all the trees gone?

When humans arrived in Australia, perhaps 60 000 to 70 000 years ago, they brought with them the great forest destroyer—managed fire. They used it to keep the forests 'clean' as well as to encourage the new growth that brought kangaroos and wallabies to graze. On the newly burned forest floor game could be tracked more easily. In Tasmania an early settler bemoaned the extinction of the full-blooded Aborigines, not from any feeling of guilt but because only they knew how to burn the forest country and keep it open for travel as well as for grazing by cattle and sheep.

In south-western Australia the pattern of the local Aboriginal people was to start fires in late spring and early summer and let them burn slowly through the forest, until the autumn rains put them out. The wheat farmers bribed them to stop this firing pattern since such a fire could spread to the wheat fields when they were ready for harvest.

In other parts of Australia this regular burning would have diminished the rainforests and converted the edge of wet forests into woodland, while the woodland changed to grassy plains with scattered trees.

Extent of existing natural forest in Australia (general outline of more extensive areas)

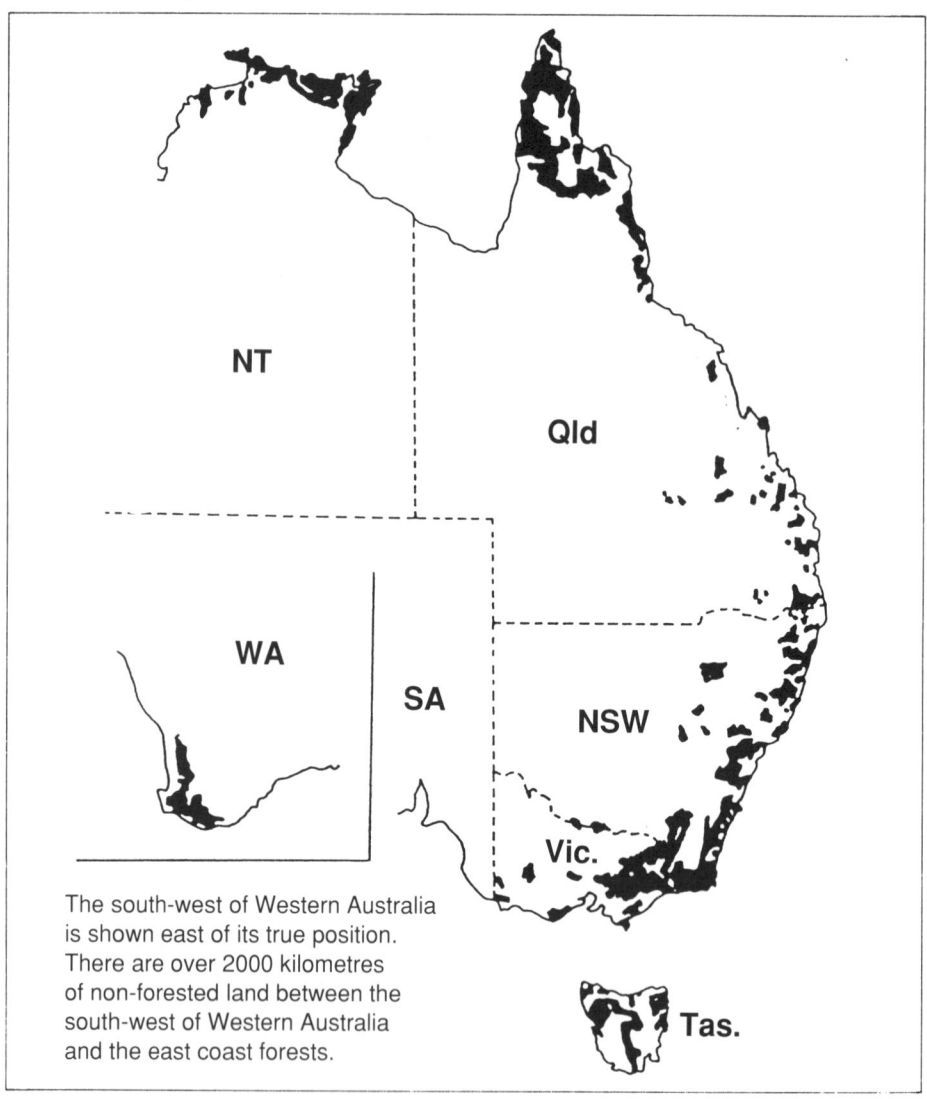

NT

Qld

WA

SA

NSW

Vic.

The south-west of Western Australia is shown east of its true position. There are over 2000 kilometres of non-forested land between the south-west of Western Australia and the east coast forests.

Tas.

Source: Compiled with the assistance of Mr Frank Bullen of the Division of National Mapping and Dr J. A. Carnehan, by reference to Division of National Mapping maps of introduced crops and pastures; land use; managed oversets; and land tenure, including nature conservation reserves, military training areas and Aboriginal lands; and Department of Arts, Heritage and Environment.

Notes: 1. Small state forests cannot be depicted at this scale, and discontinuous forest patches on private land are also omitted.

2. Even though considerable commercial timber may be produced from them, 'woodland' or 'scrub' regions with trees generally under 10 metres and/or a protective foliage cover of less then 30 per cent are not depicted on this forest map.

Cedar cutters, a colourful breed. (Forestry Commission of New South Wales)

Slash and burn

Elsewhere in the world some hunter-gatherers gradually became agriculturists. In the forests they began the technique known as 'slash and burn'. A patch of forest was cut and allowed to dry, then burned. The ash provided fertiliser for crops for a few seasons and then the people moved on to slash and burn a new area. The surrounding forest gradually healed the scar created by farming.

Australian Aborigines, though they knew of food crops from their near neighbours, the islanders and Macassans, never became farmers. Why plant food when nature provided such a bounty? When European settlers arrived at least a third of Australia was still covered with forest and woodland. The needs of agriculture meant a rapid disappearance of both.

Forest as a crop

At first timber cutting was basically a 'mining' of the forests, selecting and lopping particular timbers. Along the eastern coastline huge stands of red cedar were decimated. The first timber was exported in 1795 and a century later the industry was over.

It would be unfair to blame the cedar cutters for this. Their work provided an early capital inflow. Hard on their heels farmers finished the job, with the forest changing to farmland. It was only later that governments began to realise that our forests were a valuable resource which, if properly managed, could last for ever.

The normal pattern in those days was for foresters to go through an area to be thrown open for cutting. Trees to be felled would be marked and then the area rested for perhaps seventy years or more. Because only single trees were removed there was minor change in wildlife values. Indeed until about forty years ago state forests were the most valuable recreation areas for bushwalkers and nature lovers, as national parks were few and far between.

Ironically, modern forestry is gradually reverting to a sophisticated form of slash and burn. With economists calling the forestry tune came a demand for greater returns from the state forests. The woodchip industry started in Eden in 1968 and its success tempted more and more states to follow suit. It is now flourishing in New South Wales, Tasmania and Western Australia. Victoria is coming to the party with Queensland flirting with the use of mill wastes and plantation thinnings. Judging by the experience in other states this is the thin end of a woodchip wedge!

The basic technique is to fell completely all the trees in a section of forest known as a coupe. Most of the timber that can be used as sawlogs is taken to the mills while much of the rest is used for woodchips.

'Forests are forever' is a claim made by forestry public relations. Neither native forests, nor the abundant and varied wildlife they support, survive intensive clear-felling. Heavy equipment used in clear-felling affects over 50 per cent of the area. This is essential for the current ecologically and environmentally unjustified woodchip export industry. Such activities are based on an incorrect and narrow 'timber-production' philosophy which ignores the many other roles native forest ecosystems must play. (John Turnbulls)

Some observers claim that should the supply of woodchip timber be meagre, sawlogs are also fed into the maws of the chipping machine. This is reminiscent of the old commercial whaling days when factories needed a certain number of kills each day. If animals of the legal length were not present smaller whales were killed and authorities turned a blind eye!

'Useless' trees, old with many hollows or small or twisted, are slashed with the rest and the total litter burned. A few parent seed trees are left to provide new growth. Sometimes seeds or seedlings are sown on the ash beds. The once mixed forest becomes a monoculture as coupe succeeds coupe.

Conservationists have forced various changes such as making the coupe sizes smaller and preventing clearing along creek beds or along hilltops and the like. However, such rules are often more honoured in the breach than the observance.

Though the next cut in the growth may not take place for another 40 years inevitably such forests become stands of even-aged trees with less and less wildlife interest. It is the 'useless' trees with their hollows which are so important in keeping diversity. About a third of all bush animals need such hollows for shelter or nesting. So clear-felling in forests is just as damaging as dredging on the ocean bed!

The balance sheet

While the returns for the timber cutters are excellent, for the public they are less so. There has been no attempt to charge royalties adequate enough to give a return for the resources and to replace the trees used. Because the government sold the wood so cheaply it discouraged private enterprise from moving into the field of tree production.

The money offered by the chippers has caused the destruction of many forests on private farmland, particularly in Tasmania, and such destruction is planned for Western Australia.

Plantation forest

Even before this backdoor method of creating plantations out of native forests began, the urgent need for timber had led to plantations being started, particularly in South Australia where there were no commercial native forests. The present softwood plantations cover 1.8 per cent of the total forest area and it is expected this will gradually rise to about 3 per cent.

It is important to note that at present softwood plantations provide 26 per cent of our timber needs, although they have only a small area compared to our native forests. Our native forests produce about nine million cubic metres of timber while overseas plantations of eucalypts, with only one-third of the area of our native forests, produce some 30–40 million cubic metres.

Even with the new coupe cutting the Botany Department of the University of Tasmania has found that the proposed sustained yield from Tasmanian forests will only be 24 cubic metres per hectare per year while eucalypt plantations in other countries reach up to 60 cubic metres per hectare each year. It is obvious why foresters feel the need to increase the commercial take of timber from native forests and to increase the area of softwood plantations.

The old way: single tree felling instead of clear-felling. (Qld Forest Department)

Bushwalking in a state forest thirty-five years ago. In those days, before clear-felling, such forests were dedicated to multiple use.

Other values

The only aspect we have considered so far has been the wood produced from a forest. Yet these regions have other values, some of which rank higher than the wood produced in particular places. In earlier times, the Aborigines used forests not only for hunting and for timber, but for religious rituals. After white settlement the forests were used for grazing stock, as a source of eucalypt and other oils, as places for beekeepers to set up their hives in season, for water catchments, for recreational needs and as an educational resource.

Above all they were reservoirs for wildlife. Particular animals in some states are still hunted commercially. As gene pools they preserve our present diversity and this is particularly important in rainforests since their economic potential as a source of drugs and other chemicals is as yet unknown. If they are destroyed we may lose material of great economic value.

The Australian wild dog or dingo. Brought to Australia a few thousand years ago by Aborigines this species became a serious predator on the native wildlife. (See also p. 96)

A lily trotter runs over water plants in Kakadu National Park. Twenty years ago these birds were rare in the park as water buffaloes had destroyed most of the water plants. Many waterbirds are now returning to this region of huge wetlands. (See also p. 90)

Lord Howe Island has been described as the most beautiful island in the world. It has great scientific value as well as beauty and is one of the World Heritage Sites to be protected for all time. (See also pp. 128–9)

Wildlife

Many studies have compared wildlife values of plantations, particularly those of introduced species, compared to native forests. The lack of diversity in artificial plantations as well as the new style 'coupe cut' forests can be summarised as follows:

1. A plantation is a simple structure with only one species of tree. The disappearance of shrubs and other layers of medium-sized trees means that habitats for animals are minimal.
2. Many animals need hollows for shelter and raising young. Without such hollows they cannot inhabit forests. The ground layers of fallen branches and logs are essential for a number of species. This affects creatures as diverse as insects, frogs, reptiles, birds and mammals.
3. A plantation by its very nature provides less variety and less abundant food. This is particularly true of introduced species since they leave their insect attackers behind in their home country, at least for a time. Later, when the attackers do arrive, such monocultures may suffer much more massive attacks, since there will be no buffering effect through the diversity which provides controls on the pest species.

Other dangers

Strip mining has not been a threat to forests until recently. The most dramatic destruction was of the jarrah forests of south-western Australia when mining for bauxite began. Mining not only physically destroyed the forest but the numerous vehicles spread the spores of the cinnamon fungus which destroyed the jarrah trees and many other native species through the disease 'dieback'. The replacement forest sown after mining included species from the eastern states so the original bushland has gone forever.

Fire

The other danger is controlled burning to prevent wildfires which can be disastrous to lives and property. Again, south-western Australia offers a classic example of what can happen. Until 1953 all efforts were directed to excluding fires. This resulted in a leaf litter that accumulated for thirty to forty years without being returned to the

Pine forests and smaller monocultures can be used for recreation, such as trail bikes, four-wheel-drive vehicles, horse riding, etc. (Forest and Timber Bureau, ACT)

Farmers turn from traditional crops to growing pines. (Forestry Commission, Victoria)

soil. The result was a dangerous fire hazard. Expensive firebreaks had to be maintained.

After the devastating fires of 1949–50 the policy was changed and control burning began in 1954. This proved better from a timber point of view but the situation is complex. The spread of dieback is being linked with changes in burning patterns.

Private lands

With the rise of woodchipping, timber on privately owned land is being destroyed. This is reasonable for the farmer since it earns a cash crop and the land cleared can be used for pasture or other purposes. If governments want these trees replanted to make a sustainable crop then financial incentives will be needed.

Rainforests

Can rainforests be logged without losing their integrity? In New South Wales all the remaining rainforests have been declared national parks so they are safe. In Queensland the opinions of conservationists can best be expressed in the following statement from Dr J. Winter, a rainforest ecologist with the Queensland National Parks and Wildlife

Service: ' . . . with our current state of knowledge, the only safe management policy is to leave as much rainforest as possible intact, a course that some countries, for example, Australia, are more able to follow than others.'

Since the days of white settlement the original rainforest has been reduced by as much as 75 per cent. It is accepted by all experts that rainforests are an important genetic resource but forestry departments know that they are also an important source of timber, particularly in Queensland which mills three-quarters of the Australian production. The veneer and marine plywood industries must use local rainforest timber or import their needs.

In 1985 a working group on rainforest conservation produced a report for the federal government which provided a 'comprehensive but flexible package of measures which, if implemented, would constitute an effective attack on the problem of loss and degradation of rainforest nationally . . . ' Its success depends on the states co-operating with the federal government. At present Queensland refuses to do so. A decision by the federal government to nominate the tropical rainforests for World Heritage listing has saved a major section of such forests.

Agroforestry

This is a technique by which timber and agricultural production are carried out on the same land. It has been used in some river red gum forests and pine plantations as well as eucalypt forests. A study in Western Australia showed that agroforestry can offer substantial long-term profits.

A possible scheme for a forest farm

1. Plant rows of radiata pine into improved pasture (10 000 trees per ha).

2. Harvest hay from between rows until the trees are big enough to resist damage by stock.

3. When the trees are 5 years old, thin to 500 per ha; prune remainder to 30% of their height.

4. At 8 years, prune to 40% of height to allow more light through to the pasture and to maintain wood quality.

5. First commercial thinning occurs at 12 years–thin to 200 trees per ha, prune to 50% of height (thinnings are sold as pulpwood or fence posts).

Chipping logs for pulpwood may be carried out on the property.

6. Harvest sawlogs at 25 years–then re-establish pasture and tree crop.

Source: Ecos.

Wood yields (cu m per ha) from widely spaced radiata pines at Mungalup, WA

Age (years)	Pulp	Case logs	Saw-logs	Spacing:
12	32	77	—	8-12 years, 500 trees per ha
20	30	—	145	13-20 years, 250 trees per ha
25	—	—	300*	21-25 years, 125 trees per ha
				*Projected yield of sawlogs at final harvest in 1982

This productive site near Collie, WA, was planted in 1957. It has shown the potential of widely spaced plantations. In fact many of the sawlogs felled in the 1977 thinning were sold at premium prices for peeling into veneers.

Source: Ecos.

Urban forests

In essence this is the planting of trees in urban areas, not necessarily as a crop but mainly to provide beauty and other environmental advantages. The more urban forests resemble natural forests the better, so native species are favoured.

Many of the older cities of the world have obtained both recreational and timber values from the planting of trees in the heart of urban complexes. Basically an urban forest should be large enough so a person a hundred metres inside it will not be able to see the houses on the edge of the forest. The species should be mixed, with most but not all native. The main advantages are more pleasant air conditions and an area for passive recreation. Victoria has given a lead in this field and has a well-developed advisory system as part of their aim to 'green' the state.

Future options

It is obvious from our early history that multiple use is the best solution for forestry management. It is also obvious that our increasing timber needs will come from plantation forests. Many of our traditional rural crops are becoming unsaleable in the export market and this land could be turned over to tree farms. Native forests could still be used for selective logging and woodchip production from mill wastes but the main raw material would come from the plantations.

Already there is the problem of providing places for four-wheel-drive vehicles and trail bikes. Plantation forests are ideal for this purpose and some pine plantations in the Australian Capital Territory are being used in this way.

It is time for all state forestry Acts to be rewritten so that multiple use is a legal management requirement. The general public should be invited to comment on such management plans as is done with national parks. Both public and private forests should have comprehensive regional planning with federal input in terms of conducting regular national reviews of policies. Royalty charges must always be levied to include the full costs of new forest production as well as the value of the timber. In the federal sphere taxation incentives can be provided to encourage more farmers to begin growing trees.

The success of the Greening Australia project shows that volunteer workers are keen to assist in this act of national resource conservation.

Forests into parks

Could we turn all our remaining native forests into national parks? At present our native forests produce 10 million cubic metres of forestry products, 40 per cent as sawlogs and the rest as pulpwood. Exotic softwood plantations provide 4.5 million cubic metres annually, 65 per cent as sawlogs and the rest pulpwood. Yet these plantations cover only 2 per cent of the total forest area. Trees grown as a crop are therefore 20 times as productive as trees taken from native forests.

Present plantings should ensure that by 2020 we will produce 16 million cubic metres annually, far more than we produce today. Also, it is hoped to obtain 9 million cubic metres from our eucalypt forests. It is entirely practical to increase the area of plantation forestry to produce this 9 million, sacrificing only one-twentieth of our native forests and reserving the other nineteen-twentieths as national parks. In the long run this means we will save all the native resources for our future.

Marine resources

For a people obsessed with the sea and with almost the entire population living near the ocean, we have taken little account of our marine resources. In bringing European traditions to a new country we have not realised that this is a new sea and needs new methods of 'cultivation'.

Farming the sea, just as farming the land, must be built on a solid base of knowledge so research is vital. Stop–go funding will not work. There must be continuing research with an input of private enterprise money to take advantage of its results. Then our present Cinderella may become a princess.

Here are some comments from a recent lecture by Professor Peter Sale of the University of Sydney:

I am a marine scientist and I am concerned that Government has lost interest in the need to foster marine science . . . Beginning in the mid-seventies, the Federal Government began to increase support for the marine sciences and technologies. A number of major initiatives were achieved. The Australian Institute of Marine Science was established in Townsville and has become an important centre for tropical marine research. CSIRO's involvement in marine research expanded with the development of the new labs in Hobart and the Divison of Fisheries and Oceanography was split into two: the Divisions of Fisheries Research and of Oceanography. The oceanographic research vessel *Franklin* was commissioned as a national research facility operated by CSIRO but accessible to marine scientists from any institution. It is our first and only such vessel . . .

The Great Barrier Reef is one of the three primary destinations for our overseas tourists. Tourism in Australia is growing rapidly. Present growth is well in excess of ten per cent per annum, and present value to the economy rivals that of the mining industry. If the Great Barrier Reef is to retain its attractiveness we will require increased scientific and technological research so as to be able to manage the increased tourist impact. Potential negative impacts include degradation of the reef due to construction and operation of tourist facilities as well as that due to increased recreational fishing and fossicking on reefs . . .

Fisheries are a growing component of our primary production. In 1984/5, their value was $524m, of which $439m was exported. The rock lobster and prawn fisheries are our most important and were valued at $158m and $160m respectively in 1984/5. Recreational fisheries are also of economic value. In fact, they are far more important for Australia's economy than the commercial fisheries, because amateur fishermen are estimated to spend $1125m per year on boats, fishing gear and the like. . . Aquarium fish are worth many more dollars per kilo than conventional fishery products, and are in high demand in Bonn, London, New York and elsewhere. This industry employs 40 000 people in Sri Lanka . . .

Offshore oil and gas production was worth over $5000m in 1985, and generated substantial government revenue in the process . . .

Future marine research has much to offer Australia in economic benefits, and the stimulus given over the past several years has begun to repair decades of previous neglect. Let's all remember that in marine research we cannot import and must "buy Australian". But to do this there must be a continued awareness by government of the need to make the purchase.

Fishing and overfishing

Many groups are interested in our marine resources. All are competing for their particular interest and in this battle the fate of the resource is often overlooked. Take the tuna fishing industry. In the 1970s this was an important industry with a major catch at Ulladulla. Here and in other places tuna were caught and canned, until in the 1980s the Japanese demand for raw fish resulted in much of the take being sent, packed in ice, to make the favoured dish, sashimi.

Fishing with longline and pole and hook was a conserving method according to the New South Wales fishermen, but the South Australian purse seining took whole schools. Further west the fishermen were catching small individuals. By 1983 the bluefin had disappeared from commercial harvesting.

Not learning from experience, the fishermen then turned to the other tuna including yellowfin and bigeye. The federal government tried to bring the fishermen together with the director of the Australian Fisheries Service and the federal Minister to work out a management plan for the yellowfin industry before it went the way of the bluefin.

There are problems. Most important is a lack of knowledge of the yellowfin life history. The Ulladulla people believe it is a non-migratory population which would make controls easier. However, a small number of tagged fish indicate that part of the stock migrates along

Abalone industry faces ruin

Victoria's multi-million dollar abalone export industry is in jeopardy because of over-fishing, inadequate policing and lack of research . . .

(The *Age*, 25 March 1987)

the eastern coast from as far north as Papua New Guinea and Fiji. In these waters purse seiners are at work and, being outside territorial waters, are not under our control. With a four-year lifespan and the ability to produce many young it is hoped the yellowfin will last better than the bluefin. Japanese fishermen in the past have taken huge hauls off our coast.

Perhaps it is time we started to look at the world picture. International conventions already regulate the taking of tuna in the eastern Pacific and the Atlantic oceans. We need a similar convention to regulate the western Pacific waters so that all the nations involved in these seas can agree on a management plan which will safeguard the basic stock for the years to come.

Marine resources are much richer than the obvious creatures such as fish, oysters, scallops and algae. Our coastal zones are among our most precious environmental assets. Almost the entire population of Australia spends at least some of its recreational activities in these regions. Places such as the Gold Coast in Queensland have suffered in terms of building muddle and this lack of planning is echoed in the web of scientific and management agencies around Australia, all stubbornly holding on to their small empires. The Australian Marine Sciences and Technologies Advisory Committee is helping co-ordinate some of these agencies and in 1981 published a policy for the 1980s.

Overfishing is an obvious danger which has already caused problems in the abalone, tuna and shark fisheries, to mention only three. Yet this is not as important for our long-term future as the destruction of the estuaries and bays, essential habitats for the food chains on which so much of our marine life depends. Damage is caused not only by unwise building developments such as filling in salt flats or destroying mangroves for open space or housing but also through pollution which damages quality. An important objective should be to keep not only foreshores of estuaries and bays but the whole of the coastline under conservation management. Only essential recreational requirements should be allowed in the 400-metre land zone back from the sea.

Already in New South Wales one-third of the coastline is protected from any further urban encroachment. That is an example other states should follow.

Perhaps the Great Barrier Reef Marine Park pattern is the ideal management solution. Here all the traditional activities such as fishing, shipping and yachting continue but under management guidelines to ensure that the marine qualities remain undamaged and in many cases are improved.

Similar multiple use management has already been suggested for our forests and regional parks. The system should be extended to Bass Strait, the South Australian gulfs and offshore islands, the west coast of Western Australia with its offshore coral reefs and islands and the Gulf of Carpentaria. The marine systems of Antarctica, in which Australia has a large involvement, should be declared a world marine park.

Priority actions

A number of immediate actions are needed, some of which are listed in the NCSA document in Appendix 2 of this book. Basically we need adequate research to find out what resources we have, the dangers which threaten marine ecosystems, the legislation needed to protect these resources and emergency plans for rapid action when danger threatens. This structure is already in place for oil spills and whale rescues, to give two examples.

In some areas only governments can play a significant role but in many cases there is need for co-operation with voluntary agencies. This works well on land in bushfire control, while volunteers are the main workers in whale rescues and saving oiled seabirds. Such involvement is increasing rapidly. The Royal Australasian Ornithologists Union is carrying out work on migratory seabirds such as waders.

The community accepts the importance of bodies such as Red Cross, Meals on Wheels and the like. With changing patterns of employment and early retirement, thousands of useful workers will be available for voluntary services and wise governments will offer the small amount of seeding money which will ensure a huge return in free labour. The friends of museums, botanic gardens and national parks often have huge memberships and the volunteer work they carry out amounts to millions of dollars annually.

The Australian fishing zone

A few years ago international agreements on the rights of nations to the seas that wash their shores gave Australia a huge marine territorial region. The 200 nautical mile Australian Fishing Zone, and the rights to the continental shelf, mean that we now control an area about the same size as the land surface of our continent.

Marine resources

Our marine resources must be considered in the coastal zones which include estuaries, in the Australian Fishing Zone and continental shelf, and in our involvement with international waters since many marine resources are not entirely territorial. Migratory fish such as tuna, and birds like the shearwaters, can be affected by changes taking place away from our own zones of responsibility. Pollutants entering the oceans or atmosphere may end in faraway places: DDT used far to the north was found in penguin tissues in the Antarctic.

Australia is not a lucky country where marine resources are concerned. Our coastal waters are low in nitrogen and phosphorus. Many nutrients come into the sea from river runoff but our arid continent provides very low quantities compared to other nations. We lack any major upwellings from ocean currents so most of our resources are scanty and almost fully exploited at present. In 1980 our export trade was only $250 million, 80 per cent of which came from rock lobsters, prawns and abalone, all at the luxury end of the market. Commercial fish stocks are important mainly for the domestic market.

A non-renewable resource, our supply of oil and gas is being exploited heavily. Bass Strait has provided us with valuable materials and the more recent discoveries of north-western Australia may prove even more valuable in the future. These fossil fuels will run out so we should use such bonanzas to plan for a sustainable and prosperous future.

There are minor resources such as pearl and food oyster culture. Giant clam experiments

Mangroves are outdoor laboratories and also nurseries for much of our marine life, including commercial species such as fish, oysters and prawns. (Queensland Museum)

on the Great Barrier Reef may lead to an important marine farming industry for clam meat. Estuaries and bays may be able to provide more food, although such places already provide about a quarter of our total marine catch.

Management

The responsibilities for control of all marine resources are divided between the state and federal governments. Traditionally this was a state responsibility; the federal powers were more recently recognised. It was a commonsense decision to allow control of all freshwater fisheries and those offshore to the three nautical mile limit to remain with the states. Beyond that the control becomes a federal responsibility. At times particular regions are managed in co-operation, while the Australian Fisheries Council provides a forum where both commercial and conservation voices can be heard.

Sooty terns—in earlier times many of these colonies were destroyed by collectors taking eggs for food.

A coastal people

Painter John Olsen gave a graphic description of Australians as being inhabitants of a giant saucer, most of us clinging to the rim to prevent ourselves from sliding down the plughole of the outback! Most Australians live on the coast and it is on the shorelines and estuaries we find our recreation, not only summer swimming and surfing but all year round in terms of boating and fishing.

Often there are management problems when recreational desires conflict with professional needs. In economic terms the recreational use of such marine resources is far greater than any commercial use since most coastal towns depend almost entirely on them for their existence. A steady movement to more tropical waters is also the pattern of people all over the world.

Pollution

This steady movement to the sea creates the problem of the fouling of the waters by pollutants of all kinds. Storm water carries vast amounts of oil, pesticides and fertilisers into the sea; we still do not understand the full effect of these pollutants. The construction of jetties and buildings on dune systems has altered the ebb and flow of sand, sometimes creating beaches, often destroying them. In recent years the dangers of heavy metal pollution have been important since these tend to accumulate in marine food stocks.

Some of these pollutants are being brought under government control as discussed in the section dealing with wastes, but the two most obvious pollutants are oil spills and sewage.

Saving the gene pool

This is another way of stating the need to keep the present diversity of marine life. The establishment of marine parks and other reserves is the ideal way of preserving the present diversity. Such establishment has lagged considerably until the last ten years and is still missing in most of our coastal regions.

The first marine park in Australia was part of Bouddi National Park in New South Wales. After that promising beginning interest in such reserves languished. Various reserves to protect fish stocks and other marine life were set aside from time to time but the great leap forward came with the creation of the Great Barrier Reef Marine Park, the largest of its kind in the world. This is not a national park in the usual sense but a managed area, similar to the European regional park and rather like my suggested national heritage park since it is jointly managed by the state and federal governments.

The Seas and Submerged Land Act 1975 brought all land below low tide mark under commonwealth control and is a great help to marine conservation, at least while sympathetic federal governments are in power.

Saving the whale

Australia took a lead in world affairs with the passing of the Whale Protection Act 1981 which banned the taking or killing of whales and dolphins. The state governments also protected seals. The dugong is protected by special legislation so all marine mammals appear safe from direct attack. Aborigines and Islanders still carrying on their traditional lifestyles can hunt these animals and recent reports indicate that this needs policing, to prevent commercial exploitation. Marine reptiles such as turtles and crocodiles are also protected.

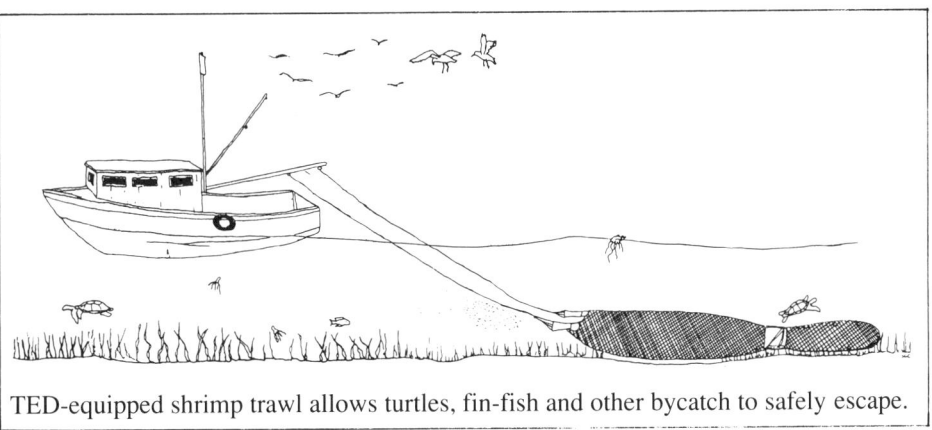

TED-equipped shrimp trawl allows turtles, fin-fish and other bycatch to safely escape.

Source: Centre for Environmental Education, Washington.

7. International conservation strategies

Foreign aid

An IUCN report in 1982 stated: 'The intensity and frequency of human disasters is often related to the ecological mismanagement and especially to that of vegetation through unwise clearing, burning and overgrazing.'

The purpose of the World Conservation Strategy was to prevent such ecological misuse. Bandaid treatments are no help and our foreign aid would be better spent, for example, on the huge eucalypt planting campaign once carried out in Ethiopia but now cut due to a change in the flow of cash for such assistance. It is imperative that all our foreign aid should have to pass through the sieve of the World Conservation Strategy to ensure that we do not by our help create worse problems in the future.

When our heartstrings are tugged it is easy to be like the unwise parent who, softened by a child's fear of the dentist, supplies painkillers to stop the toothache. The decaying tooth does not go away. It strikes with greater force later when it is too late to do anything but take drastic action. Most agencies have accepted the principle that aid must be planned to provide a cure not a palliative.

Population control

The World Conservation Strategy mentioned that this problem may need a separate strategy treatment. It was a watershed when in 1983 IUCN joined forces with the International Planned Parenthood Federation with a membership of 116 nations 'in their efforts to protect the global environment and to reduce the pressure of population growth'.

Forget the cry that says 'the world has plenty of food, why can't we send it where it's needed?'. If the standard of living of the present world population were raised to our level, all the surplus would vanish. Literally there are not enough resources of food and other materials to fill its needs.

International treaties and conventions

Several treaties and conventions dealing with conservation principles and policies are already in place. The major ones will be discussed in the sections to follow. The NCSA statement on internation co-operation is:

a)   Strengthen consultative arrangements and information exchange between the Commonwealth and other government and non-government bodies concerning Australia's participation in international conservation agreements and programs.
b)   Ensure that the objectives arising out of the NCSA and WCS are taken fully into account in Australia's dealings with other countries.

c) Promote international understanding of the importance of the unique physical character and ecology of Antarctica and seek appropriate forms of management for the continent.

United Nations Environmental Programme (UNEP)

It is fashionable to downgrade the work of the United Nations. Any huge organisation will have faults but rarely do the critics look at the excellent initiatives it has organised, even though some have not lived up to the first high hopes.

UNEP was established as the result of the exciting conference on the human environment held in Stockholm in 1972. For many of the general public this was the first airing of environmental problems on a world scale.

UNEP was planned to have three major interests. One was to monitor what was happening around the globe as an environmental watchdog. Another was co-operation with other UN bodies so that environmental aspects could be taken into account in all plans for the future. Finally there was to be direct help in particular programmes around the world.

Unfortunately UNEP has become less and less significant on the world scene mainly because it has never had the status of a United Nations specialised agency. Nations, especially powerful ones, do not like international interference in their affairs, no matter what damage they are doing. It is an attitude familiar to all Australians: some state governments resent federal interference in their 'right' to ruin the environment in pursuit of quick returns, or money or job production.

Another problem is that the Stockholm Conference and the present headquarters of UNEP in Kenya are all in the northern hemisphere. For example, 5 June was chosen for World Environment Day as an ideal time for the celebration of the environment. The delegates forgot the other half of the world which had the opposite season. Had they suggested 1 December as the day, commonsense would have rejected the idea in favour of a time when local climates would allow celebrations to be held out of doors. If the whole world had been considered World Environment Day would have been held either in spring or autumn, both suitable in either hemisphere. This minor matter highlights the tunnel vision in many international agencies which forget we are one earth and tend to concentrate on the more powerful parts of the globe.

The United Nations General Assembly should examine all the machinery for handling environmental matters and see what needs to be improved, rather than dismantle the present organisation. Even more important is that public servants should appreciate that not all conservation wisdom is enshrined in government organisations. The voluntary conservation movement is both skilled and ready to play its part in decision making.

One of the problems for nature conservation had been the parrot cry of federal governments of the 1950s and 1960s that nature conservation was all a matter for the states. In the early 1970s this attitude changed dramatically. Today it is taken for granted that the Australian government has an important role in both the national and international field, particularly in regard to treaties and conventions. In the early years of federation most of these dealt with military matters but with the spread of treaties to civil and political rights and then the environment, the scene has changed considerably.

One of the problems is that successive federal governments decided that before Australia becomes a party to such a convention the agreement of all the states should be obtained. It may take up to a decade for progress to be made. This need not be a bad thing; planning that matures slowly may have advantages. Yet in some cases agreement may come too late to save the resource. The international conventions to which Australia is a party give an idea of the immense potential power available to any concerned government interest in taking care of our environmental future.

Convention of International Trade in Endangered Species of Wild Fauna and Flora (CITES)

This convention was finalised in the United States in 1973 and Australia had both signed and ratified the agreement by 1976. More than 91 countries have now accepted this accord.

The history of trade in wildlife is an ignoble one over most of the world and since the first white settlement Australia has suffered from an uncaring rape of our wildlife. The story of the near extinction of our koalas for the fur trade is well documented. The local extinction of the elephant seal and the near extinction of our fur seals took place in the early days of settlement while the koala slaughter ended only in the 1920s.

Since then controls on export of our wildlife have become progressively stricter, although unfortunately the illegal trade still occurs. Rare bird species such as the golden-shouldered parrot suffer. Trade in the skins of both the saltwater and freshwater crocodile brought their numbers to dangerously low levels, but this has now ceased.

The world's rarest reptile, the south-western short-necked swamp tortoise, now numbers only a few dozen and, although protected in reserves, is not safe from poaching. Such an animal is easily carried in a coat pocket through Customs. It is reported that these tortoises have been sold for $5000 in Europe so it is obvious that those taking part in this revolting trade consider the risk to be small and the reward high.

The trade in cagebirds is better documented. For every bird reaching its destination alive it has been estimated that a hundred may die on the path from the first netting in the wild to the final destination.

The Australian government has recently introduced legislation to control these threats to our wildlife. Individual states now have export permits dealing with endangered species in an attempt to monitor what is happening inside Australia, but until every state has an adequate environmental policy a steady drain on our rare wildlife will continue. Animals trapped in one state are often illegally exported to another. No matter whether these individuals live or die, there is an overall loss for that species.

It has been suggested from time to time, usually by farmers suffering some local loss from attacks by birds on their crops, that an export trade in our commonly regarded pest species would not only bring relief but also make economic sense. This belief is naive. The overseas demand for species such as galahs and corellas is at artificially high prices because the export is illegal. Once a few thousand birds are legally trapped and exported the prices would crash. Losses in the more common species would soon be made up by breeding, so the next season would see the same numbers as before, and farmers would get no relief in the long run.

The rare south-western short-necked swamp tortoise whose numbers have been reduced to less than fifty individuals. Loss of habitat was the main cause of destruction but smugglers could make it extinct. One animal can be hidden in a coat pocket and later sold for thousands of dollars.

There is also the ethical problem of catching creatures used to the wild, then caging them for the rest of their lives. Most conservationists accept the cagebird trade in aviary-bred individuals. They would react strongly to any attempt to trap wild individuals.

The present position is that all countries ratifying CITES stipulate that government permits are required for all trade in endangered or vulnerable wildlife. Appendix 1 lists the endangered and Appendix 2 lists the vulnerable species. The species listed in Appendix 1 may enter trade from the wild only in exceptional circumstances and may not be imported for a primarily commercial purpose unless the specimens are captive-bred or artificially propagated. Appendix 2 animals and plants may enter trade from the wild but this is monitored and carefully controlled to ensure that vulnerable species do not become endangered. Any nation may ban all trade in its wildlife, whether common or endangered.

Governments are now controlling this problem, but non-government organisations have provided much support. The International Union for the Conservation of Nature and Natural Resources (IUCN), an organisation with both governmental and non-governmental members, provided support for the secretariat that was initially funded by the United Nations Environmental Programme (UNEP). The Species Survival Commission, a body formed by IUCN, identifies the endangered and vulnerable species of wildlife around the world. Experts volunteer for this task and a number of committees provide the information.

IUCN, the International Council for Bird Preservation (ICBP), and a host of national groups are also always alert for breaches. Among the international non-government organisations is one colloquially known as TRAFFIC, Trade Records Analysis of Flora and Fauna in Commerce. This group is working with governments and gradually bringing order into the present wildlife trade as well as warning of problems in the future. World Wildlife Fund Australia has now accepted the financial burden of TRAFFIC in Australia.

As with all laws, only an aware and concerned public can make them effective. In particular, tourists buying products containing skins, feathers or horns must ensure they are not from any prohibited species. Customs will not allow entry of these products. Even more important for those involved in commercial imports is to know how to identify the prohibited species. Many firms have lost large sums buying shoes, handbags and other manufactured material, only to find them refused entry into Australia. The Curator of Reptiles at the South Australian Museum has devised a simple training booklet which enables anyone to identify all the species on the prohibited lists after only a few days' training.

The most important step needed in Australia now is to make sure that the habitats of all endangered and vulnerable species are protected by secure reserves. These need to be well managed to prevent poaching.

Convention for the Protection of the World Cultural and Natural heritage

The conference that produced the draft for this convention was held in Paris in 1972 and more than 85 countries have now ratified the convention. Australia had ratified it by 1975 and the effects were to be far-reaching.

A number of conditions must be met to earn a place on this prestigious World Heritage List. Areas of outstanding natural value must also be of a high scientific and aesthetic quality. Australia now has seven items on the list with a nomination of the tropical rainforests of Queensland to be considered in December 1988. The seven areas are stages I and II of Kakadu National Park, the Great Barrier Reef Marine Park, the Willandra Lakes region, the Lord Howe Island Group, the Western Tasmanian Wilderness National Parks, the New South Wales Rainforest National Parks and Uluru.

Nominations for inclusion on the list are the responsibility of the commonwealth government, although usually this responsibility is effected in consultation with the relevant state or territory. Nominations are forwarded by the Department of Foreign Affairs

The Franklin River—the conservation battle to save the wilderness helped defeat the conservative government of the time. Increasingly voters are regarding environmental issues as important in deciding how they vote. (Tasmanian National Parks and Wildlife Service)

to the World Heritage Committee for consideration. Once a nomination is accepted it becomes a federal responsibility to ensure that each area is handled under the following criteria—identification, protection, conservation, preservation and transmission to future generations.

It was this responsibility that made the Western Tasmanian Wilderness National Parks region so critical. The massive change to the Franklin River by a proposed new dam was a breach of the convention. The state government had changed its political colour and wished to cancel a contract into which it had willingly entered, a move which would be refused under common law standards. The government of the time offered the Tasmanian government massive financial support for an alternative scheme. The state government refused. The federal government refused to take overriding action as it was legally and morally bound to do under the convention. The opposition Labor Party promised that if elected it would carry out the terms of this international agreement.

After a vigorous campaign, in which environmental issues played a major part, the Labor Party won government and proceeded to make good its promise. A High Court challenge was defeated and so the powers of international conventions became part of Australian law. With one bold stroke Australia regained the lead in one international field and earned the thanks of conservationists around the world.

Usage of World Heritage Areas and other national parks

Of the civilian population aged fifteen years and over, it is estimated that 790 000 persons or about 7 per cent visited the World Heritage Areas in Australia in the twelve months to April 1986; the most frequently visited being the Great Barrier Reef which was visited by 436 000 persons.

In the same period it is estimated that about 35 per cent of the civilian population aged fifteen years and over had visited a national park in Australia (other than a World Heritage Area).

(Australian Bureau of Statistics, 1986)

The role of the Australian Heritage Commission was most important. This commission is a unique authority established in 1975. Seven part-time commissioners and a small research and secretarial service carry out its duties which, in simple terms, are to list 'the things we want to keep'. The 'things' may be any item in the natural and built environment.

Since 1975 about 8000 items have been put on the Australian Heritage Register. Although protection of such places is mainly the duty of the state governments, avoidance of unnecessary damage or change is binding on all federal departments. Fortunately in most states there has been co-operation between both state and federal governments to protect items on the register.

In the Franklin River decisions the commission played a vital role negotiating for over four years in order to convince the state governments concerned that the first five items should go forward for nomination to the World Heritage List. And it was its presence on the list that eventually saved the Franklin River from destruction.

Convention on the Law of the Sea

This was an initiative taken by the United Nations in 1967. It had a long and troubled passage before major acceptance, and some nations, the United States in particular, still have reservations about its provisions.

The United Nations had always accepted that the seabed is the common heritage of all mankind but the ramifications of the final convention are vast. There are four main thrusts of importance to Australia.

The first is that the territorial zone which was three miles (the distance the old-fashioned cannon could shoot a cannonball) has been extended to 12 nautical miles. The second is to develop more measures to protect the marine environment, and control fishing so it can continue as an exploitable, renewable resource. The third is to bring the seabed under international control. Lastly, a new concept has been developed. Exclusive Economic Zones (EEZ), which extend the territorial zone to 200 nautical miles, have been declared. By this change our marine territory has almost doubled. There is an obligation on us however to use this huge area or license others to do so. Although it is one of the world's largest fishing zones, it is not particularly rich.

It took many years of conservation lobbying to force the government to control the Taiwanese pirates who raped our northern seabeds of giant clams and other marine wealth. Similar kid-glove treatment of Indonesian fishermen continues and they have been given permission to obtain fresh water from Ashmore and Cartier Reefs. This supplies an excuse for their harvesting of protected seabirds and their eggs. They also hunt endangered species such as dugongs and dolphins as well as make semi-permanent camps on some islands, so breaching our quarantine security and offering the opportunity for the introduction of dangerous diseases to Australia.

Laws are meaningless unless policed, and we need a comprehensive system of coastal and marine surveillance, using not only paid staff but also volunteers. Some steps have been taken but there is need for more action. Perhaps our greatest need is for the navy to have more smaller vessels based in northern Australia, not in Fremantle, Sydney and Jervis Bay!

International Convention for the Regulation of Whaling

Australia was an original signatory to this convention which came into force in 1946.

Agreement between Australia and Japan for the Protection of Migratory Birds and Birds in Danger of Extinction and their Environment

This interesting policy developed through concern for the future of the Japanese snipe, a bird which was hunted on its arrival in Australia and whose breeding grounds were under threat in its home territory of Japan.

There has been a wider concern about the conservation of migratory species of wild animals which could include the present agreement on birds. Unfortunately, for no good reason, Australia has not acceded to this convention. This should be treated as urgent.

A similar agreement was signed with the People's Republic of China in 1986.

Approximate annual migration route of the Bass Strait muttonbird

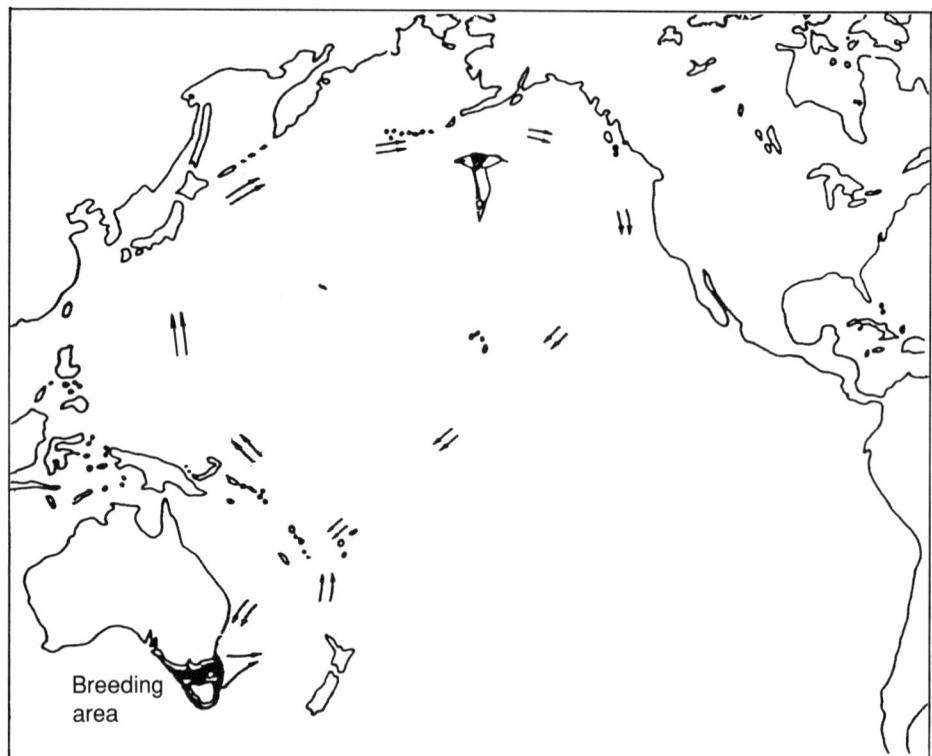

Breeding
area

After leaving the breeding burrow on offshore islands of southern and south-eastern Australia, these birds follow a 35 000 kilometre route through the Pacific each year and return to the same breeding place. Behaviour patterns such as this pose many questions.

Convention on Wetlands

The first wetlands conference, an attempt to stem the worldwide destruction of these habitats, took place in 1972 at Ramsar in Iran. The convention has been accepted by forty nations. Australia became the first contracting party when on 8 May 1974 it signed the convention without reservation to ratification. Twenty-seven wetlands, with a total area of 1 294 090 hectares, have been designated by Australia for inclusion on the List of Wetlands of International Importance.

Oil pollution at sea

This century has seen a new threat to the environment in terms of oil pollution, caused partly by the immense amount of fuel being carried on tankers across the world and partly by

The approximate annual migration route of the little stint

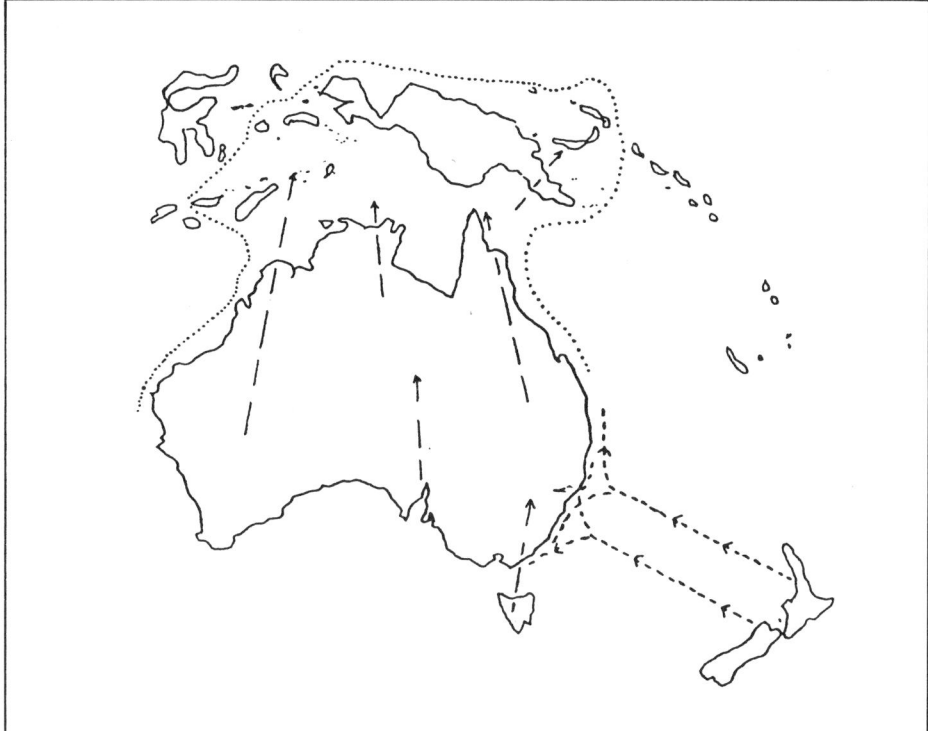

Dots indicate the general path of the little stint, a tiny wading bird that breeds in eastern Siberia and western Alaska and spends the rest of the year (our summer) on the estuaries and mud-flats of Malaysia and Australasia.

runoff into the sea of oil spilled on land. Every year about two million tonnes is spilled into the sea from ships, not only through wrecks but also by 'accidents' caused either by apathy or by deliberate contempt for international laws. Another four million tonnes of oil pours into the sea from other sources.

Some damage caused by massive spills is obvious; tens of thousands of seabirds are killed or injured and washed ashore with oil-clogged feathers. More insidious and long-lasting is the damage to various marine food chains and the killing of exposed reef life through oil slicks or continuous pollution.

A number of conventions have been drawn up in an attempt to hold the pollution in check. Particularly important is the need to prevent pollution resulting from the cleaning of ships' fuel tanks at sea. This would stop the deliberate discharge of oil at sea as a cheap way to solve this problem.

Convention on the Conservation of Nature in the South Pacific

There has been co-operation at federal level with the countries of the south-west Pacific but an important step forward could be taken if Australia acceded to the Convention on the Conservation of Nature in the South Pacific prepared in 1976. At present only Samoa, New Guinea and France have signed this convention.

Although aimed at the development of national parks in this region, the convention does have wider nature conservation purposes. One article deals with the establishment of a list of species of wildlife threatened with extinction. Another deals with the interchange and training of personnel in the conservation of nature.

Once the convention comes into force the first step would be to establish a secretariat to be funded by the party nations. We are well placed with the Australian National Parks and Wildlife Service already trained and able to play a key role in implementing the convention. We need a similar convention for the Indian Ocean.

The International Tropical Timber Organisation (ITTO)

The world's only international organisation aiming to conserve tropical forests by reforming the timber trade met for the first time in Japan in March 1987. The organisation is committed to encouraging the development of national policies aimed at sustainable use and conservation of tropical forests and their genetic resources and to maintain the ecological balance in the regions concerned. Australia is a member of this organisation. Australian firms are involved in rainforest logging and we import tropical timber.

Antarctica

'Great God. This is an awful place.' So wrote the British explorer Robert Falcon Scott, who died with his companions on the return from the South Pole.

Here is a continent half as large again as the United States, an ice desert drier than Australia, and paradoxically the world's greatest reservoir of fresh water, held bound as an ice sheet with a depth of 1.6 kilometres, in some places as thick as 4.5 kilometres. It may be an awful place for humans but it is a paradise for penguins and seabirds, whales and seals. The surrounding oceans teem with life. The key member is a crustacean known as krill.

This seven-centimetre-long shrimp provides the food supply, directly or indirectly, for much of the region's wildlife. The actual mass of krill is not known with certainty but estimates range from 800 million to 5000 million tonnes. Many biologists believe we could harvest an annual catch of 50–60 million tonnes, roughly the same as the present annual world fish catch. Obviously krill is a tempting prize.

Antarctica most probably has rich deposits of minerals including that most valuable of all, oil. Other aspects of this frozen land are also important. If all the ice sheets melted the world seas would rise between 45 and 90 metres. Such a rise would flood most of our agriculturally important lands and also the most populous cities. This vast mass of ice is also important in regulating world climates.

Yet Antarctica is far away and it is often forgotten that this is one earth. Dr Kenton Miller, at that time the director of IUCN, on his return from Antarctica, said his most lasting impression was of a large graph on a wall which measured the carbon dioxide content of the atmosphere. The line showed a steady rise down the years with each annual cycle marked by a tiny rise caused by the pouring of carbon dioxide into the air as a result of the use of oil, coal and wood in the northern hemisphere each winter. The pollution from such fires spreads around the earth with frightening speed.

It was discovered earlier that the pesticide DDT, used in the northern hemisphere, quickly found its way into the Antarctic. It was detected in the tissues of the penguins of the region.

The problems of Antarctica are many, but a few examples will highlight some of the questions facing Australia politically since we lay claim to the largest part of this southern continent.

The slaughter of the great whales, when killing went almost unrestricted in terms of management, left a gap into which other animals moved. Food chains in these icy seas are far simpler than elsewhere. Basically many of the whales, seabirds, fish and squid eat krill or some other predator which eats krill. The krill themselves feed on the vast swarms of drifting microscopic plants called phytoplankton which grow rapidly during the short southern summer.

When humans harvested only the larger animals at the end of such a food chain the others were left unscathed. When we hunt further down the chain towards the food base, the effects on all the larger forms of wildlife can be catastrophic.

The removal of the great whales gave smaller animals such as seals and penguins an advantage. The smaller minke whale relished the new conditions and the ample food caused its age of maturity to drop from fourteen to seven years while the numbers jumped. Crabeater seals, which eat krill not crabs, exploded in population while chinstrap and gentoo penguins also increased. Even in Australia we have had an increasing number of Antarctic penguins and other sea creatures being stranded on our southern shores.

Today, ignoring the lesson of the destruction of the great whales, some nations are already harvesting krill. There has been little public outcry as, while the killing of whales can be seen to be both savage and cruel, there is little sympathy for a lowly shrimp. Little knowledge of its life history means that properly managed harvesting is impossible. No protection is offered to those swarms that occur near the great penguin nesting areas. The common fish of the region are also being harvested in the same way.

Here is an obvious recipe for potential disaster. What can Australia do? We lay claim to 30 per cent of the undisputed land, with New Zealand taking 14 per cent, Norway 18 per cent and the United Kingdom, France and Chile asking for very modest percentages. Yet these are only paper claims as the United States, Soviet Union and most other nations refuse to recognise the validity of any territorial claims.

While we say we 'own' some 6.5 million square kilometres and have made many important geographical and scientific discoveries in our region, we should do what we can to ensure the conservation safety of the area while accepting that this vast land and its seas are part of the common heritage of humankind.

Such commonsense may be half a century away, and the de facto controllers are the signatories to the Antarctic Treaty of 1959. This set up a policy on all the lands south of the 60 degrees parallel of latitude. Like all artificial borders it ignores the biological fact that the

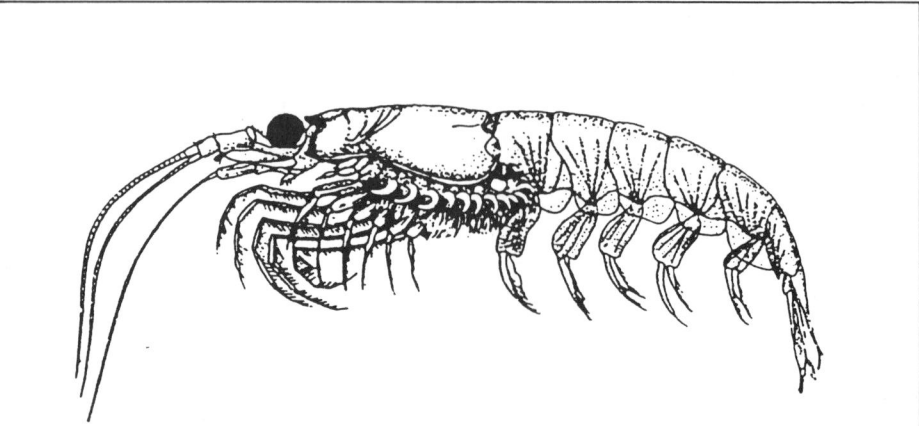

Krill, *Euphausia superba,* is a shrimp-like crustacean, 7.5 centimetres long overall. This animal is a vital link in the food chains of the Antarctic on which depend penguins, seals and the great whales.

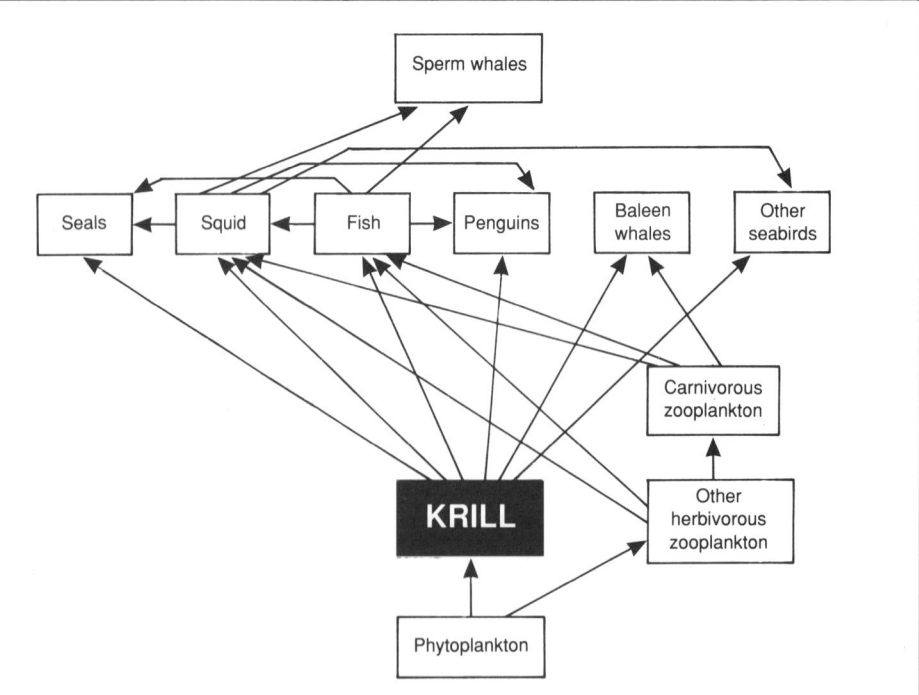

Krill form the central link in the Antarctic food web—both at sea and (via the seabirds and the nutrients their droppings bring to the nesting sites) on the land.

dividing line should be the Antarctic convergence where the cold polar-ice water sinks below the warmer northern waters with massive effects on the local wildlife.

The treaty nations have agreed on a number of useful conservation measures. These include using the land for peaceful purposes only, a free exchange of scientific information, and the banning of all nuclear material. The first eight signatory nations have been swollen by fifteen more which take part on a non-consultative status.

It is often claimed that this co-operation is a diplomatic wonder. The explanation is that so far there has been no economic reason for disagreement. Where whales were concerned the organisation proved a paper tiger, just as it is in the case of krill and fish harvesting. When one acquisitive nation wants to start drilling for oil or mining in Antarctica, the treaty will prove powerless.

Immediate action should be to draw the rest of the world into the management of this region, at least in consultative terms. The World Conservation Strategy has already outlined the objectives needed for this area.

In accepting these goals the Australian government should push for the whole of the marine ecosystem to be managed along the lines of the Great Barrier Reef Marine Park. This means that some economic activities can continue, although only after management guidelines have been prepared. Some areas need to be zoned as national parks, both on land and at sea. The Antarctic Treaty organisation is already in place and working reasonably well so it should continue as the present de facto authority.

There should be annual reports to the United Nations, outlining the present position and putting forward plans for the future. In this international forum all nations could take part in the debate and recommend amendments, such decisions having the force of international authority.

To 'keep the treaty partners honest', the United Nations should set up a watchdog committee. Ideally such a monitoring body could consist of UNEP, IUCN and the Scientific Committee for Antarctic Research (SCAR).

As the years pass a shift of legal authority to the United Nations would gradually evolve but the management would be left to those with the greatest skills. The United Nations would step in only if any dangerous conservation action was planned. Whatever present feelings are about the United Nations, commonsense indicates that only with a strengthening of this organisation can the world proceed to a sound future, not only in terms of conservation but also of world peace.

This is part of an evolutionary conservation process. Just as nature conservation is still mainly a matter for local organisations with the federal government taking a more supervisory role, so the gradual shift from a few nations in charge of the Antarctic to world control of this common heritage can be regarded as inevitable.

8. The future

Creative writers have often expressed human needs much earlier than economists or ecologists. Long before the multiple value of wetlands was recognised the poet Gerard Manley Hopkins wrote:

> *What would the world be once bereft,*
> *Of wet and wildness? Let them be left,*
> *O let them be left, wildness and wet,*
> *Long live the weeds and the wilderness yet.*

The great nature philosopher Henry David Thoreau would have welcomed the Franklin decisions, mourned over the Daintree rainforest destruction and despised those who see the Kakadu wetland only as a place to be raped of its natural wealth in favour of uranium mining.

So much for fine words. Where do we go from here? Creating the NCSA was the first big hurdle; yet it is only the beginning. A small committee of state and federal government representatives, manufacturers, timber firms and farmers as well as conservationists have made a number of recommendations which are listed in the following chapter. Few of them have been carried out.

State–federal co-operation

It is essential that all governments, federal, state and regional, ensure their policies and activities are in keeping with the NCSA objectives. All the statutory and other authorities, including departments of forestry, soil conservation and the like, should, where necessary, rewrite their objectives in terms of the strategy. One example is the New South Wales Royal Botanic Gardens and Herbarium Trust Act, which does not mention nature conservation. Good work has been carried out in this field in spite of the official policy but it can be stopped if an unsympathetic Minister comes into power.

The national priority actions in the NCSA should be examined so that areas of responsiblity can be decided. In some cases these must be dealt with on a national or broad regional basis. The Murray–Darling salting and erosion problems come into this category.

One worthwhile initiative dealing with gene pool preservation has been the co-operation between state and federal governments in managing national parks which are of substantial significance, although not quite worthy of World Heritage nomination. Some are now being cared for with federal government expertise and money to assist the state in preserving the present diversity. Other priority actions can be handled at state or regional levels with the relevant bodies developing their own timetable for tackling them. The various Ministerial Councils need to be aware of long-term aims. A federal–state Ministerial Council also needs to be set up to oversee co-operation.

National Advisory Council

Above all we need a National Advisory Council as a high level forum for consultation and a source of expert advice to all branches of government. Such a council could be similar in organisation to the present consultative committee with a non-government chairperson. It must be able to make public statements without any of the constraints placed on government servants. All Cabinet submissions at both state and federal level should include a paragraph where appropriate noting how the suggestions fit the NCSA guidelines.

What applies at these high levels should continue down the stream of command to shire councils and other groups planning policy. Even farmers or home-owners can develop a conservation policy for their own future. The following chapter gives a few ideas on the role of individuals.

Promotion

This is most important. The highest expertise should be used to make people aware of the importance of conservation planning for the next 200 years. Perhaps the best promotion of all is to take action on highly visible problems and show how they are being solved. The Murray River is a wonderful example of what might be done.

The Consultative Committee outlined such a plan in some detail. Projects in which federal, state and regional bodies and even individuals, could be involved were detailed in their report as follows:

1. Tree plantings along key sections of riverbanks for erosion control.
2. Plant regeneration on selected groundwater recharge areas to reduce salting. Trees act as water pumps and keep saline water tables well below the surface.
3. The same techiniques should be used on paddocks destroyed by salt.
4. Instead of pouring the two valuable resources of sewage and fresh water into the river, they could be used to irrigate tree farms. A success in this field would stimulate other communities to develop their own projects.

What is happening in the states at present?
Tasmania and Queensland

Neither government has endorsed the strategy because of their opposition to any intrusion of federal advice into their state affairs. Even worse, neither is planning to develop its own state strategy.

South Australia

In many ways this state leads the field. The principles of the NCSA are used to determine the outlook of all government departments, while Cabinet submissions must take the strategy into account. There is co-operation with the federal government in the joint management of what have been called National Heritage Parks in this book. South Australia is preparing its own state strategy through the Environmental Protection Council.

New South Wales

This government is not convinced of the need to develop a state strategy but it has set up a committee to oversee and report on what is happening in the state. Conservation groups here are determined to press for the development of a state strategy.

Western Australia

A strategy has been developed.

Victoria

This state has gone further along the road and developed a State Conservation Strategy which is a valuable document as it is an operational plan. It could serve as a model for all the other states.

Regional and sectional strategies

This is good in parts, but not as far advanced as it should be. The Standing Committee on Forestry of the Australian Forestry Council has produced a National Forestry Strategy.

Other regional and local strategies are being developed and we hope a web of them will cover Australia. Dr Geoff Mosley, in a report to a follow-up conference in Ottowa in 1986 which looked at what is happening around the world, commented that Australia is regarded as one of the leaders in this field: 'the conservation strategy in Australia has taken only a few steps of its long march but already useful experience has been gained about the best way to travel'.

It would be good to know that Australia which, in the final years of the last century and the first decade of this, led the world in so many ways was now once more offering new vistas. The federal government under Gough Whitlam was a watershed for the environmental movement and this work was continued by Malcolm Fraser. It is unfortunate that care of the environment policies, judging by the Opposition attitude to mining in Kakadu National Park in particular and our other national parks in general, no longer unite politicians of all parties.

At the Third World Wilderness Congress in Scotland Mr Barry Cohen, the Minister responsible for the environment, was given a standing ovation by the delegates, a first for any Congress. He was reporting on the Franklin story. It would be tragic if at a future World Wilderness Conference an Australian should have to ask for a minute's silence to mourn the loss of a World Heritage site destroyed by greed and stupidity!

The basis of *Saving Australia* and achieving the objectives of the NCSA is to integrate conservation and development, to emphasise their interdependence and common ground. The next stage of NCSA development must be more precise. We must face the ecological dilemma of how to integrate the needs of an expanding human population and community expectations on how we can improve living standards, within the framework of limited natural resources.

Our need for greater agricultural production places demands on our natural resources which they are often unable to tolerate. Australia's lifestyle may already be as high in terms of material things as our life support systems and productive ecosystems can sustain. We need a more equal distribution of material goods and an expansion of values to be satisfied. A walk in the bush uses no material resources and can be repeated again and again. The same applies to the use of libraries, going to the theatre, recreational fishing, swimming, playing cricket, football and a host of other activities. For most people such activities lead to a richer lifestyle without any increase in the possession of material objects.

We must also remember the ecological realities detailed in the second chapter of this book. Although the NCSA has received reasonably wide acceptance in Australia the problems of carrying out its programmes are difficult, though not impossible. It will require the adoption of a new perspective and a multidisciplinary approach to land management.

For a junior Ministry, such as Arts, Sport, Environment, Tourism and Territories, to ensure this approach is almost impossible. A high level commitment to proceed from one of the Cabinet sub-committees is needed. The Economic Planning and Advisory Council

A survey: Should government spending be increased, decreased, or kept the same?

	Increased (%)	*Decreased (%)*	*Kept same (%)*
Health programs for the elderly	78	2	18
The environment	73	5	19
Aid to the homeless	71	5	21
Health services for the poor	71	5	22
Nutrition programs for mothers and infants	55	6	34
Reducing acid-rain pollution	54	11	25
Low-and moderate-income housing	54	11	32
Loans and grants to college students	54	15	29
The food-stamp program	33	24	36
The space program	33	27	34
The military	31	25	38
Star Wars	23	35	26

Source: Time magazine, 30 March 1987.

(EPAC) in the Prime Minister's Department should endorse the concept of the strategy and actively support its adoption by other departments, governments and sectors of the community. Taxation regulations should be altered as part of the national strategy to achieve national goals which incorporate the NCSA rather than as a means of raising revenues for unstated purposes. National goals which set down long-term overall intentions or aspirations have first to be agreed.

'Greenways' of Australia

The linking of all our present reserves with 'greenways' would enhance the survival of all our wildlife as well as making 'green roads' available for people.

About ten years ago I put forward the idea of planting 'koala corridors'. My suggestion was that interested people should plant koala food trees along roadsides and through paddocks and other open spaces, so that an ambitious koala could travel along such a green road from Townsville south and west until it reached South Australia! The idea caught the public imagination and is underway.

Behind my interest in such corridors was the knowledge that these strips of trees and shrubs would assist not only koalas, but many other kinds of plants and animals that find even narrow belts of vegetation of value.

Perhaps now is the time to expand the scheme and create a web of 'greenways' across Australia. It is not always realised that the 5 per cent of our country reserved as national parks is made up of 'islands' in a sea of human change caused by agriculture, mining and urban expansion.

As every naturalist knows, island populations are vulnerable to extinction, with the dodo of Mauritius the most famous example. Nearer home the Lord Howe Island woodhen was saved in the nick of time. The linking of all our present reserves with 'greenways' would enhance the survival of all our wildlife.

Equally important is the fact that Australians are among the most urban people in the world. The wildlife most of us can see and enjoy every day is the kind that flourishes in small patches of green such as road edges and urban parks. Some states have roadside verge committees to safeguard these green corridors, both in city and country, although the Northern Territory, Queensland, New South Wales and Tasmania still ignore this important aspect of nature conservation.

A few years ago the President of the United States set up the President's Commission on Americans Outdoors. Its brief was to make a two-year study on how Americans viewed their own natural areas and how many were able to enjoy them. Among the conclusions of the study was the fact that 84 per cent of Americans liked walking for pleasure and wanted their green places to be close at hand.

The commission therefore recommended that communities across the nation should establish more 'greenways'. Such plantings should be on both private and public land and enhance lakes and other waterways as well as beaches. The aim was to provide people with access to open spaces close to where they lived, and to link the rural and urban green spaces in the American landscape.

Also, with the predicted greenhouse effect causing climatic change, such corridors would allow plant and animal species to move north or south to areas of suitable climate, either wetter and warmer or drier and cooler, depending on their needs.

Some states, including Western Australia and Victoria, have begun planting such corridors. The federal government should assist in making surveys so that we can create a giant green web of nature, spun across our continent. The greenway web would link small and large reserves, lonely beaches, busy urban parks, lakes crowded with yachts and all the natural landscapes of Australia.

National goals

The following proposals by the Australian Science and Technology Council (ASTEC) might serve as a basis for a set of national goals:

Social cohesion   A harmonious society without cleavage along ethnic, religious or class lines and with provisions for peaceful change in politics and instructions;

Community welfare   A provision for community health, both mental and physical, care of the disadvantaged and disabled, opportunity for educational achievement;

Cultural and scientific advances   A high quality of creative effort in the arts, sciences and technologies and a wider understanding of them.

Next there should be a set of specific statements about what needs to be achieved by certain dates. These objectives should be stated exactly and be subject to measurement and evaluation.

How can we achieve the following?

- a stable or decreasing national population should levels rise to non-sustainable numbers;
- disincentives for large families;
- full employment;
- technology to perform the boring, repetitive tasks;
- growth in the economy in areas where this does not result in an increase in using non-renewable resources;
- management of living resources to improve their quality or quantity and reverse past errors;
- action plans in taxation policy;
- disincentives for use of non-renewable resources;
- incentives for land development and construction industries to concentrate on maintenance and improvements of existing cities and facilities;
- incentives for employment in service industries such as tourism, leisure, domestic services and the professions;
- royalties and taxes that reflect the availability of resources;
- income tax reduction to encourage employment;
- royalties that reflect the limited availability of natural resources of land, water, minerals and forests.

Instead of allowing populations and consumer demands to increase without question we need to be much more aware of ecological limits. National policy on population stability, resource conservation and environmental protection should be developed by all government bodies and integrated as vital aspects of national economic policy. Those in the community who realise this need should join with others to act as a powerful lobby group. If this is done Australians will have a firm basis for moving towards the next 200 years.

Public education

Most states have developed curricula which allow teaching of environmental facts to primary and secondary students. Many universities and other tertiary institutions have courses leading to careers in the field of environmental work.

It cannot be stated too often and too strongly that Australia needs a well-educated population to solve our many problems, both urban and rural. Education is the real wealth of any country and if we are a lucky country it is because we have enough natural resources to keep us solvent while we build up a bank of education skills. We have a long way to go before we catch up with other more advanced countries in terms of the proportion of our people who have some form of tertiary education, the advanced skills we need so urgently.

Media

Schools are only one factor, and not the major one, in educating the community in environmental commonsense. Television is the most persuasive of all media with radio and newspapers not far behind, especially in creating values. It is not important if the private media outlets are owned by twenty millionaires or only one. In either case the ownership is usually unsympathetic to any political interference with business.

While the media were generally supportive of the conservation movement in its early days, the situation has now changed. Lobbying by pressure groups is influencing political decisions and putting multi-million dollar development proposals at risk. It is becoming more and more difficult to obtain media coverage for protection of both the natural and the built environment. This is why there must always be a substantial section of the media in public ownership, or laws to force media to offer similar coverage for answers to blatant enviromental misrepresentations.

From a study of the past conservationists can learn how to do battle in the future. The following stories show how small, medium-sized and national groups can achieve important gains for the environment.

Kellys Bush

This tiny area of 7.7 hectares of bushland forming part of the foreshores of Sydney Harbour was zoned as 'open space'. To everyone's surprise, a developer bought the land and put forward a plan for its development as a housing site. The local council approved this change from bush to bungalows. It was a smooth takeover, found everywhere in Australia and always defended on the grounds of 'progress'. This usually means progress in terms of private profit, but a step back where the public is concerned.

The residents already had a Hunters Hill Trust which had managed to keep something of the old quality of this small suburb. It is a historic section of old Sydney, with grace and elegance.

Thirteen housewives decided to save the bush in which they had played as children. They formed the 'Battlers for Kellys Bush' and many laughed at their simplicity in trying to keep back the bulldozers. The Battlers fought grimly even though the position looked hopeless. The then Premier had promised action before the election but forgot the promise once back in power. Then came a suggestion that the Battlers should turn to the unions. Over most of

Underwater on the Great Barrier Reef. This coral wonderland was saved from lime mining and oil drilling by public pressure, aided by Queensland unions. It is another of the World Heritage Sites and a tourist asset for Australia. (Valerie and Ron Taylor) (See also pp. 128–30)

A saltwater crocodile and a great egret in Kakadu National Park. Egrets were under threat many years ago when they were hunted for their feathers. The crocodiles were hunted much more recently for their skins, until both the freshwater and saltwater species became endangered. Today they are protected. It is hoped crocodile farming will allow the reptiles to become important as a supplier of skins, while animals in the wild remain tourist attractions. (See also pp. 126–8)

Sulphur-crested white cockatoos. These large birds are common in forest country of eastern and northern Australia and in some places are agricultural pests. Overseas, such birds fetch high prices. Since the Australian government does not allow the sale overseas of Australian wildlife there is an illegal trade in smuggling our more interesting animals, particularly members of the parrot family. (See also pp. 126–8)

Frill-necked lizards. These harmless dragon lizards became prime targets for smugglers when a craze for keeping them swept Japan a few years ago. (See also pp. 126–8)

Education is a vital tool in conservation.

the world the trade unions had been derided as being incapable of caring for the finer qualities of life.

With some trepidation the Battlers contacted the New South Wales Builders Labourers Federation. This branch had unusual leaders with a social conscience, common in the earlier days of the movement. The union had decided that if substantial conservation groups considered a group of buildings or a piece of bushland should be preserved they would take action to save it. Stalemate!

The bush had an uneasy peace until a change of government brought in a new Premier and party which decided to make environmental concern a major aim. Finally Kellys Bush became part of the parklands of Sydney Harbour.

More importantly, the action by the Builders Labourers Federation passed into popular language as the Green Ban. It spread to other areas and The Rocks were saved. Soon after, trade unions effectively protected the Great Barrier Reef from oil drilling and mining. This Blue Ban forced an enquiry which led to the creation of the Great Barrier Reef Marine Park despite the opposition of the Queensland government.

Today some Australian unions show an environmental concern superior to many other community groups. It is proposed that in Kellys Bush a tablet will be erected commemorating the world's first Green Ban. Even more important, changes were made to state laws so that normally law-abiding citizens were no longer forced to take illegal action simply because the law offered no forum in which their views could be heard.

Organ Pipes National Park

This 65 hectare park, 20 kilometres from Melbourne, was declared in 1972. Its main feature was a collection of basalt columns resembling organ pipes. Ninety per cent of the rest of the area was covered with weeds, including boxthorn and thistles. The Friends of the Organ Pipes began a 15-year campaign to regenerate the plants. Other volunteers helped, including rock climbers who removed weeds from the organ pipes. It has been estimated that more than two million dollars of voluntary restoration work has been carried out, bringing back the old beauty to a weedy wasteland.

Kings Park

On the other side of the continent was another bushland region. The 400 hectares of Kings Park near Perth, Western Australia, was first reserved in 1870; down the years it became larger and a showplace of Perth.

There had been various suggestions for hotels and hospitals to be sited in this green space. Then came a dangerous and stronger applicant, the Perth City Council, who decided this was the only possible place for an Olympic swimming pool.

The Womens Service Guild, led by an energetic and grand old battler, theosophist and one-time suffragette, Mrs B. M. Rischbeith, began the fight. It was joined by other groups and soon a private members' bill on a non-party vote won the day, much to the chagrin of the government.

Swan River

Having saved Kings Park the group turned to save the Swan River from being gradually reclaimed for carparks and freeways. This battle was lost. So Perth, which once lay as a gem below Kings Park edged by the silver thread of the Swan River, was in one stroke condemned to mediocrity. The beautiful reflection pool of the river became a mess of roads and carparks.

This struggle highlighted another conservation truism—no battle is ever entirely won until the gains have been enshrined in legislation.

South-west Tasmania

This national battle began with the fight to save a small lake in the heart of south-west Tasmania. Lake Pedder with its kilometre-long scalloped sand beach along the eastern shore was, during summer months, the goal of many bushwalkers and naturalists.

The Hydro-electric Commission, which had dominated the economic planning of Tasmania for many years, wanted to flood the lake for another dam to create energy for the companies which came to this state because power was cheap. The economic balance sheet showed a number of capital intensive metal refining industries employing little labour but much energy, leaving behind a loss represented by massive pollution, both of land and water. Dams steadily engulfed the beauty of the high country.

Conservationists from all over Australia came to fight the plan to destroy Lake Pedder. I met one of the leaders, Ron Brown, and many members of the conservation group he had helped organise. We all agreed with Thoreau that 'a lake is the landscape's most beautiful and expressive feature. It is earth's eye; looking into which the beholder measures the depth of his own nature'. Despite all our attempts and the assistance of the federal government, Lake Pedder was blinded in 1972!

The Duke of Edinburgh, a keen conservationist, wrote: 'Many people believed that the decision to flood Lake Pedder was a mistake . . . I hope that never again will Australians have cause to question so vehemently a decision on any conservation cause.'

He should have known that developer leopards need time to change their spots. The commission, greedy for more power, proposed to create a new storage scheme and this time the majestic wild river that ran so exultantly free in its narrow gorge was to be destroyed.

But now the conservation movement was stronger and better organised. It had won many battles, the most important of which was the creation of the Australian Heritage Commission which led to the declaration of five key national parks as World Heritage areas. As a commissioner I had helped with the four years of patient lobbying necessary to get agreement from the states to nominate these regions. We knew the power of the World Heritage Convention and we knew, from the Lake Pedder fight, that without that power a federal government, no matter how sympathetic, could do little.

Building on the Lake Pedder issue, a new leader, Bob Brown, had helped form the

Australian Heritage Commission

The decade has been both stimulating and frustrating. The highlight has been the establishment of the Register of the National Estate (now containing some 8000 places across Australia) and the development of procedures under which these places may be afforded recognition and protection.

We have witnessed the setting up of heritage legislation in the states of South Australia, New South Wales and Victoria and moves towards it in most of the states and territories. It is interesting to note that the areas of current federal–state conflict on National Estate matters are generally where no heritage legislation exists.

The reality after ten years of operation is that some significant places have been lost and others are still under serious threat. Some of the hard-earned gains for conservation are not assured. New pressures, quite unforeseeable ten years ago, are emerging. We are not just looking at threats to places so far officially unidentified, but a different range of threats to National Estate places whose integrity we had assumed was secured for posterity. The commission is becoming more conscious of the fact that 'wins' are but temporary and losses to the National Estate are forever.

(Bruce Davis, Chairman, Tenth Annual Report of the Australian Heritage Commission, 1985–86)

Tasmanian Wilderness Society. Soon he became a national leader. The misty loveliness of the Franklin Valley became a familiar sight on television screens around Australia and in other parts of the world. Malcolm Fraser, who had created a new policy among conservative political parties, found the pressure from developers too much and capitulated. His government would not use the powers it had under the World Heritage Convention to save the Franklin. The baton he dropped was picked up eagerly by the Labor Party whose leader, Bob Hawke, stated bluntly that if his party were elected to government, the Franklin would be saved. The rest is history.

The conservation movement had learned its lesson. Small battles can be won by small groups but the major struggles need unity. The 400 000 members of conservation and natural history societies acted as communicators to the rest of Australians.

All these battles are samples of struggles taking place over the continent and indeed in most parts of the world. The slow awakening of the environmental movement has educated the public into the fact that so much beauty, so much interest, so much diversity, so much of our economic future is disappearing.

A personal view of 2188

For I dipt into the future, far as human eye could see,
Saw the Vision of the world, and all the wonder that could be;
Saw the heavens fill with commerce, argosies of magic sails,
Pilots of the purple twilight, dropping down with costly bales;
Heard the heavens filled with shouting, and there rained a ghastly dew
From the nations' airy navies grappling in the central blue.
Till the war drums throbbed no longer and the battle flags were furled
In the Parliament of man, the Federation of the world.

Alfred, Lord Tennyson, 'Locksley Hall', 1842

I have two visions of Australia 200 years on. Both are possible. In the first scenario we will continue the story of the last 200 years, when apathy, ignorance and greed produced the environmental disasters we face today.

The Australia of 2188 would see our lakes, rivers and estuaries so polluted that they would be biologically dead. Our forests would be woodchipped into even-aged stands of single kinds of trees with as much wildlife interest as a paddock of wheat. Brisbane, Sydney, Melbourne and Adelaide would merge into one 'slurb' stretching from Port Augusta to Noosa, with mini-Gold Coasts strangling the beaches.

The last 200 years has seen us bring 10 per cent of plants as well as 15 native mammals, one bird and many other less noticeable animals either to extinction or near to it. If this slaughter continues, such attractive creatures as the numbat, the noisy scrub-bird, the lumbering northern hairy-nosed wombat, the south-western short-necked tortoise and a host of other animals, together with one-quarter of our plants, will become extinct.

The earth would become warmer as carbon dioxide levels rise due to the burning of wood, oil and coal. A greenhouse effect would then prevent heat from the earth escaping

Will the Australian coastline from Melbourne to Noosa, Queensland, become one giant 'slurb' such as Bondi in New South Wales? (Douglass Baglin)

into outer space and rising temperatures would melt continental icecaps and cause climatic changes in the important agricultural regions of the world. This would flood the earth's heavily populated regions and destroy the majority of our productive lands.

The second, more optimistic, scenario is the one I personally accept. The last 20 years has seen an amazing, almost revolutionary change in our thinking, brought about by conservationists educating the general public. Many of the trends of the last 200 years have been reversed. Twenty years ago only 1 per cent of Australia's natural genetic diversity was protected in national parks and similar reserves. Today the figure is more than 5 per cent and by 2188 it should be at least 10 per cent.

At least half our coastline will be reserved so our beaches and dunes will be in public ownership with natural bush forming a background of beauty. Cities will no longer be allowed to grow unchecked. Increased pollution, fuel costs and a rise in crime will cause a town planning revolution. People will not travel like lemmings into the social death of megapolis. Decentralisation will allow most of us to live within walking or cycling distance of our place of work. Private cars will survive only for leisure use, for which they are ideal. The slurb will be halted and cities of about one million people will provide the kind of cultural achievement most could enjoy.

Ninety per cent of Australians will be urban dwellers but they will live in a new kind of suburb. Pockets of bushland will enable wildlife to come and go, so enlivening home gardens. National parks and other large reserves with a diversity of space, and farms which will provide more greenery, will surround the suburbs allowing a quick escape from city pressures. Farmers will learn to live with the wild while roadside verges as well as private plantings will create wildlife corridors to link national park 'islands'.

Integrated pest management will save money and stop the poisoning of our waters, soils and air. Solar and fusion energy will prevent the increase of carbon dioxide due to burning fossil fuels.

The Australian population will stabilise at between 22 and 60 million depending on how much food we export.

We will have solved the problem of lack of water as 34 per cent of this resource is in Tasmania, 44 per cent in the north and only 34 per cent in the southern Australian mainland where most Australians live. New techniques will include the use of water currently wasted in sewage.

We will have enough fertiliser for farming use while our reserves of coal and gas will still be ample to see us through the next 200 years.

Though we live on an island we will help other countries to live within their land and water resources. We will accept migrants with the skills we need so our present ethnic mix will continue. Aborigines will provide a much larger measure of cultural input than at present. We will all regard ourselves as Australians, no longer hankering after other lands to call 'home'.

Our education system will be different. Slow learners and quick learners will be given the individual tutoring which will increase our resource of brain power.

We will become a conserving society, though this will not mean a drop in standards of living. They will actually increase exponentially as we find we need fewer material things but much greater services of all kinds, helping contribute to full employment. We will use all the extra labour freed by the technological society for enriching the lives of all, assisting the mentally disturbed and the care of the old.

Above all, skilled ecological management will see the land of Australia restored to the quality of 1788, with increased diversity because of our better understanding of what our natural heritage was like before the arrival of any humans.

Which path will we take?

9. Do you care for Australia?

Our resources are not unlimited. They must be conserved. Real progress is in improving the quality of our living, rather than the quantity of things we have. Expanding populations, the needs they have and the wastes they produce, mean that our earth is facing dangers on every front—air, water and land. Australia does not yet face the dangers of overpopulation, just too many people living in too few cities.

How much water do I use?

The average household in Sydney, the Illawarra and the Blue Mountains uses about 300 kilolitres of water per year (1 kilolitre = 1000 litres = 220 gallons). But many households use much more.

The average consumption is equivalent to 270 litres per person per day.

The following estimates of typical water use patterns may help you understand how much water you use for different purposes:

Toilet:	13 litres per flush.
Bath:	50 to 120 litres (half full).
Shower:	40 to 250 litres for the average 8 minute shower.
Dishwashing by hand:	18 litres per wash.
Dishwasher:	20 to 90 litres per wash.
Clothes washing:	Large automatic machine 110 to 265 litres per load, twin tub 40 litres per load.
Garbage disposal unit:	30 litres per day.
Handbasin:	5 litres.
Tap running while cleaning teeth:	5 litres.
Drinking, cooking and household cleaning:	8 litres per person per day.
Garden sprinkler:	Up to 1000 litres per day.
Car washing with a hose:	100 to 300 litres.
Filling swimming pool:	Usually at least 20 000 to 40 000 litres.
Dripping tap:	Very slight drip 30 litres per day, fine stream 700 litres per day.
Leaking pipe:	300 litres per day from a 1.5 mm hole.

Your home

- A properly oriented house with correct width of eaves warms us by winter sun and shades us from summer heat. This cuts down use of power for heating, cooling and lighting. Insulation is also valuable.
- Do not waste water. Mend leaking taps.
- Change your toilet cistern to a full-flush/half-flush type. To reduce the capacity of an existing cistern lower a brick gently into it. Take shorter showers.
- Switch off lights when not in use.
- Use soap rather than detergents. Brighter than bright in your wash may mean blacker than black in your local creeks and rivers. Check that your detergent is 'soft' or 'biodegradable'. Also make sure it is low in phosphates. This information should be on the packets or bottles. Remember, whiter than white is only an advertising gimmick. White does not really mean clean or pure.
- Tell shopkeepers you do not want useless packaging of cardboard, plastic etc. It not only prevents you examining the goods, but costs you more. It makes more rubbish to be disposed, and uses up resources of timber etc. Make the assistants remove unwanted packaging before you take delivery.
- Demand returnable containers. Make sure you recycle glass, paper, metal cans, plastic, oil and compostable food waste.
- Food with skin blemishes may be better than unblemished fruit. Farmers can cut down on sprays if you show commonsense on this matter.
- Make sure your fly and mosquito sprays are only killing insects. Pyrethrum is a safe spray. Better be sure than sorry with poisons.

Your garden

Nature gets along very well without sprays, yet there are few gardens more beautiful than bushland in spring. More and more home owners are keeping the natural bush when building a house and so have a ready-made garden.

Enrichment with other flowers and shrubs can go on steadily. Such a garden will have some insect attack but this is all part of the web of life. Damaged leaves and flowers will be repaid by the presence of birds.

The major problem is caused by planting only a few species such as roses or similar flowers. With greater diversity will come less insect attack.

Cleanliness in terms of making sure no gutters or tins hold water, that septic tanks are properly screened, will help get rid of mosquitoes.

Pests

An animal pest is only a creature in too great numbers. A few snails are not a major problem. One or two sap-sucking bugs are a point of natural history interest. Action need only be taken when pests are in excess. Small groups of aphis for example normally disappear owing to a change in the weather, bird attack, or by insect attack, especially ladybirds.

Weeds are plants out of place. Quite often the beauty of English country roads are due to plants classed as weeds in Australia. A few weeds in the garden or lawn can be tolerated.

Using weedicides in such situations may be inexpensive at first, yet finally increase, rather than decrease the problem.

Pesticides are designed to kill. This applies to humans as well. Many people have been killed by carelessness in the use of pesticides. The long-term effects are still being argued. A commonly used weedkiller, 245T after three decades of use, has been found to have a cancer forming potential. It took us three hundred years to find what a killer cigarettes are. It is criminal to risk our own lives and those of our families by using pesticides in a home garden designed for pleasure.

- Go native. Keep the natural bush on your block if you can. Australian plants, wisely selected, need no watering, once established. Most home gardens are great users of water. Native plants are just as attractive and will also bring birds to your garden which will keep insect pests in check.
- Plant trees to shade you from the summer sun. Do not burn rubbish. Make a compost heap and use the leaves again in your garden.
- Ask your local council to establish recycling facilities.
- Many pesticides kill animals directly, and being long-lived, they keep on killing along the food chain. Insects or other small animals with the poison in their bodies pass it on to the animals that eat them, and the pesticide finally reaches sufficient proportions to kill the larger animals, which may well be one of the gardener's best friends in keeping pests in check.
- Try not to use pesticides or herbicides. Waste poison must not be poured into the sewerage system or local creeks and rivers. Check with your local council where you can get rid of them.
- Do not use any pesticide which does not have clearly marked all it contains. Never buy fertilisers or soils with insecticides already mixed.
- Press your government to make sure that, by law, everything sold for the garden must be clearly marked with what it contains.
- Do not use any mercury compounds.
- Do not use long-lived poisons. These accumulate in various animals and possibly in humans. Some common kinds are Aldrin, Chlordane, DDT, Dieldrin, Endrin, Heptachlor and Lindane.
- Do not purchase any pesticide where the label recommends any form of protective equipment, even gloves—look for a less poisonous material.
- Do not use a pesticide if you are not sure what is causing your garden troubles. The trouble may be due to a soil deficiency.
- If you must use a pesticide, remember it *is* a poison:
  1. Read the instructions carefully and obey them.
  2. Wash both utensils and hands after use.
  3. Pets and young children cannot read. Make sure you have a locked poison cupboard.
  4. Be very careful of aquariums, fish ponds and natural streams when spraying and never spray with children nearby.
  5. Do not spray when plants are flowering, as useful insects such as bees will also be killed.

6. Be a good neighbour. Your neighbours will certainly not want your poison drifting into their gardens.

7. Empty containers are not really empty. In country areas, good incinerators may be used to burn all traces. In cities, municipal garbage services may be used.

8. Always keep pesticides in their original containers. Never put a pesticide in an empty food or drink container of any kind. This is a major cause of death from pesticides.

- Handpicking can be used if caterpillars or other insects are large. Soapy water will remove some pests.
- Here are some specific suggestions for getting rid of garden pests:

  Aphis—soapy water, derris spray, pyrethrum, malathion.
  Caterpillars and other chewing insects—derris dust, pyrethrum, malathion, trichlophon.
  Flies—pyrethrum.
  Sucking bugs, scale, woolly aphid—carbaryl, malathion, tar oil.
  Mites, including red spider—tar oil, kelthene, tetradiforn.
  Slug control—metaldehyde and methocarb. Baits should be covered with a flower pot to keep out of reach of children and birds. A smooth-sided container with some honey (or beer) in the bottom, will attract slugs.
  Rodents—breakback traps, in situations where birds cannot be caught.
  Bud and leaf protection—anthraquinone. Quassia chips can be bought from a chemist. Boil 7 grams in 4.5 litres of water for two hours. Dilute the yellow liquid when cool with five parts of water and use as an all-purpose spray to discourage possums, etc. Nicotine is an all-purpose poison for insects. Use 125 grams of cigarettes boiled in 4.5 litres of water for half an hour and dilute with four parts of water to make the spray.
  Damping off of seedlings—copper.
  Leaf spots rust—zineb.
  Downy mildew—zineb.
  Powdery mildew—dinocap, lime sulphur.

Your car

- Use cars as little as possible and switch off the engine when not in use. Demand an efficient public transport system. Walk. Organise a car-less day in your district.
- Make sure your car is fitted with anti-pollution devices. Keep it well tuned. Remember extra power means you may kill yourself in a high speed accident, as well as providing extra pollutants to kill the environment.
- Always carry a litter bag in your car, and try to leave every place you visit at least as clean as when you arrived. If possible, leave it cleaner.
- Buy petrol from stations which offer rubbish disposal bins and free litter bags. Also, when buying anything, whether petrol, food, etc., go to shops, factories or service stations which are not visual horrors. Tell the owners why you find their places ugly and suggest improvements.

Your district

- Demand that your local authority insists that public buildings, housing developments, factories, etc., are properly planted with trees, shrubs and open space; that your district has some bushland reserves for education and passive recreation; that foreshores are protected.
- Ask your local authority to make laws against pollution by dogs, littering, noise and advertising signs. Make tape recordings of noisy equipment in public places and try to have them played at council meetings.
- Help conduct regular drives to gather paper, bottles and other material which can be reused. Help with volunteer groups in 'cleanathons' rather than 'walkathons'. Remove signs scrawled on rocks and bridges, etc.
- Make sure your school has a well landscaped garden. Donate conservation books to its library. Suggest that your school holds an Earth Day. Has it a conservation society? Do the children belong to the Gould League?
- Native wildlife, trees and plants are what make Australia a uniquely interesting country. Prevent thoughtless and useless damage and destruction in the name of 'development'. Help start camera clubs instead of gun clubs. Plant native, not imported trees. Native animals are useful pest controls as well as adding to your district attractions for tourists.
- Get your local council to develop a conservation plan. Tree preservation should be reinforced. Use voluntary committees to give expert advice.

Your state

- Discuss conservation with your local councillors and members of parliament. Watch to see they take action. Ask them to introduce adequate anti-pollution laws. Remember pollution is caused by ignorance, greed, apathy and sheer affluence. When worried about particular items, write to the people responsible with copies of replies sent to politicians for action if needed.
- Do what you can to save roadside verges. Ask for ecologists to be appointed to Main Roads Boards, Water Boards, etc.
- Encourage adequate national parks and nature reserves, with professional management. The minimum is 5 per cent of all habitats to be preserved and 10 per cent of all coastlines.
- Lobby for your state to develop a State Conservation Strategy. Make sure that all relevant authorities, eg Forestry Departments, write the strategy guidelines into their statements of purpose. Press for all state Cabinet submissions to have a section showing how the new policy fits conservation strategy guidelines.

The federal government

- Make sure all federal parties follow the same guidelines as suggested for the states. Strengthen the federal departments involved with the environment and in particular the Australian National Parks and Wildlife Service and the Australian Heritage Commision.
- Demand that the federal government plays its part in solving worldwide problems such as ozone depletion, greenhouse effect, oil pollution, pesticides etc. and in the efficient management of international resources such as whales, seals, forests and krill.

Global

- Unrestricted population growth must be curbed or our earth faces disaster.
- Nuclear disarmament by all countries is essential. The earth can die slowly from conservation mistakes or rapidly in a nuclear war followed by a nuclear 'winter'.
- The real enemy is indifference. Make sure you are well informed by joining local, state and national organisations that are active in conservation. Set up local committees to prepare conservation plans for your own neighbourhood. Subscribe to magazines, buy books and educate your children by example in the facts of conservation.
- The Australian Conservation Foundation, 672B Glenferrie Road, Hawthorn, Victoria 3122 has published details of every conservation and natural history group in Australia.
- There is an environment centre in most of the major cities in Australia. The telephone book will give the address. Visit its office, discover how you can join in the broad sweep of activities available in your city and state.
- Tens of thousands of people are 'friends' of museums of all kinds: natural history, maritime, historical, agricultural and art, as well as zoological and botanic gardens and national trusts.

A word for conservationists

For more than forty years I have been in the thick of conservation battles. I have been interested in politics as well as natural history. Watching the struggle for existence of processionary caterpillars, conservation groups and political parties, I found they had much in common.

Edgar Snow's book *Red Star over China* told the story of the Chinese communists' rise to power. The lesson they learned was that educational work must begin among the peasants, not in the cities, because the numbers were in the country. Kellys Bush showed that the feared 'red necks' were the conservationists' powerful friends since all people want a basic quality of life.

Conservative Catholic economists made me ponder the 'small is beautiful' concept long before conservationist writer, E. S. Schumacher, educated the conservation movement on the same philosophy. I had watched my father and his friends fight the bushfire moving towards our farm by attacking the flanks. The Bradley sisters, while wandering in the bushland of Ashton Park, trying to stem the invading waves of weeds, hit on the deceptively simple Bradley method of attacking the flanks of weeds, now the accepted technique of getting rid of these alien invaders of natural beauty. Such memories moved in my mind, when priests and others gathered with conservationists to save Colong Caves. The army grab at farmlands in Orange-Bathurst and Cobar, an ill-conceived attempt to create a citadel within easy helicopter distance of Canberra, was defeated by an alliance of farmers and greenies.

All these experiences I encapsulated in the Serventy Environmental Principle. Put simply, it states: 'Never attack a major problem head-on. Attack the fringes, nibble at the edges, until the central power becomes isolated and collapses.'

Strength comes from multitudes of small groups springing to the defence of local causes that can unite on the national issues. On this firm base, district or regional conservation councils can spread their net wider. Conservation Councils are now found in most states.

What is still lacking is a Conservation Council of Australia. We need now a national Council which not only has the core of the State Conservation Councils, but also contains all the major national groups such as the Australian Conservation Foundation, the Australian Council of National Trusts, the National Parks Association, World Wildlife Fund Australia and others with memberships in thousands. Smaller groups can use their influence through the State Councils but would make a federal body unwieldy.

Working steadily year in year out, conservation groups can use the old method of 'divide and conquer'. This works whether you are tackling soil erosion, bushfires or developers keen to create a rim of Gold Coasts around Australia; those wanting to turn Australia into an empty quarry surrounded by an oil slick; urban sprawl changing green fields into carparks; or farmers trying to make paddocks into houses of battery animals.

The full weight of a Conservation Council of Australia would represent over half a million adult members. This would be a powerful weapon. But never forget that it was the thirteen housewives at Kellys Bush who breached the solid wall of developers ravishing Australia.

To this cause I dedicate the Australian Conservation Dream:

We have a dream of a world where we accept we are part of nature, not above it.
Of a world where we live within our natural resources so they can be sustained forever.
Of a world where we use our finite resources carefully with concern for the future.
Of a world where we accept a natural balance, using the energy from the sun for our major needs.
Of a world where we adjust populations in every country to the numbers the earth can nourish in dignity and with the opportunity for the fullest development of all.
Of a world where we repair the mistakes of the past to create new beauty in those regions laid waste by ignorance, greed or apathy.
Of a world where we accept that though we inherited this earth from our parents, we also accept we hold it in trust for our children.
This is our dream.

Appendix 1

An Environmental Bill of Rights

All persons are equal in dignity and have the right to life, liberty and security, promised the Universal Declaration of Human Rights adopted by the United Nations General Assembly on 10 December 1948. A Brave New World seemed to be ahead. Forty years later we know better.

Humans are a part of nature and we ignore this truth at our peril. What does human dignity or the right to life mean for those Africans caught in the Sahel region where unwise management has spread the desert, bringing starvation to millions? What did liberty mean to the Japanese killed by mercury poisoning through industrial pollution on the seas from which they obtained their food? Nearer home, what does the right to security have for workers dying of asbestosis?

As Rachel Carson warned us shortly after the Declaration of the Bill of Human Rights, the environment was also calling for care. The use of persistent poisons as short cuts to agricultural wealth meant that the world might not die in the 'Big Bang' of nuclear war but slowly in the whimper of pollution, or the effects of destruction of the ozone layer.

The time is ripe for a new Declaration leading to an Environmental Bill of Rights. The first clause must be that all people have the right to an environment adequate for their health and well-being. A number of other rights must flow from this to ensure that it happens.

In 1987 a United Nations body, the World Commission on Environment and Development, presented its report to the United Nations. The chairperson, Gro Harlem Brundtland, Prime Minister of Norway, described the report as 'a global agenda for change'. It proposed 'long-term' environmental strategies for achieving sustainable development by the year 2000 and beyond'. A major recommendation was that the 'General Assembly commit itself to preparing a universal Declaration and later a Convention on environmental protection and sustainable development'. The commission's legal experts drew up the principles such a convention should contain.

You may think me an alarmist when I say the need is urgent, but consider a few facts. Chemicals represent about 10 per cent of world trade in terms of value. Some 70 000–80 000 chemicals are now on the market—and therefore in the environment. There is a steady increase, as between 1000 and 2000 new chemicals come into the market each year. For many there is no satisfactory testing. The best estimate was that only 18 per cent of all drugs and only 10 per cent of all pesticides have been properly tested for the health hazards they may produce.

To give some idea of the dangers, 500 chemicals have already been banned, or their use restricted, in the country which first made them. What happens in the rest of the world depends on how environmentally conscious that nation is. It is no wonder that New Zealand is working hard to capture world food markets by marketing itself as a nation which grows its food without the use of poisons.

There are other problems. Outstanding worries for our future are the greenhouse effect which may change our climate; smog and other pollution caused by industries in urban areas; the slow reduction of the ozone layer; acid rain; the risks of nuclear reactor accidents and nuclear war.

What of extinction of plants and animals? In Madagascar it has been estimated that of the 12 000 plant and 190 000 animal species once living there, half have already become extinct. And the future food bowls of the world? Almost 30 per cent of the planet's land is suffering from the spread of deserts. These drylands provide the food for 850 million people. Of these, 230 million people are on soils suffering from severe effects of bad management. Exploding populations are a major cause of the destruction of the land. Sending food aid is like putting a bandaid on a deep-seated wound!

I could list dozens of other disasters, many taking place near or at home. We too have our spreading deserts and the added danger of the salting of fertile soils. So serious is this problem that about half of our land needs greater care than we are giving it at present.

Even if we solve our own problems, we cannot escape the fact that this is one world! The greenhouse effect causing a change of climate would mean a rise in sea level that would destroy urban regions and fertile soil. DDT has been found in the bodies of Antarctic penguins, another example of the mistakes of the rest of the world come to roost in unlikely places. We must not only put Australia's national environmental house in order, we must also play our part in making the United Nations environmental house safe. Australia can play a vital role. We are already highly regarded as one of the world's environmental leaders. Because of the 'romance of distance' we are accepted with affection as a nation not involved in international troubles so we will be listened to without any aura of old hates.

I hope that in the United Nations our leaders will press for an immediate start on the preparation of a universal Declaration. As soon as possible after it has been adopted they should press for the preparation of the Convention which will give it teeth. In the last century we led the world in many fields. In this century, care for the environment can be our contribution to world peace and security!

Appendix 2
A National Conservation Strategy for Australia[1]
Living Resource Conservation for Sustainable Development

Introduction

1.　　Development and conservation are but different expressions of the one process. Together they are the means of providing for the needs of the present and the future. Further, many people believe that Australians have obligations to other living things and that activities must at times be modified to respect the natural cycles of other life forms and their ecosystems.

2.　　The continuous growth of human populations and the scale of their impact on the global environment make it imperative that a new sense of responsibility be accepted if the earth's essential ecological processes and life support systems are not to be threatened.

3.　　Growth in economic activity for the enhancement of the quality of human life and, in particular, the generation of employment can be obtained without continuous growth in the use of resources by the more appropriate use of these resources.

4.　　The purpose of the National Conservation Strategy for Australia (NCSA) is to provide nationally agreed guidelines for the use of living resources by Australians so that the reasonable needs and aspirations of society can be sustained in perpetuity.

5.　　The Strategy concentrates on living resources and as such is a first step in developing the framework for conservation in Australia. To be fully effective it needs to be complemented by strategies on other subjects including energy, population and national development and by strategies at state and regional level.

---

1. *Proposed by a conference held in Canberra in June 1983. This proposal was accepted by the Australian Government. It is reproduced here courtesy of the Department of Arts, Heritage and Environment and the Australian Government Publishing Service.*

6.   The National Strategy stems from the World Conservation Strategy (wcs) published in 1980, the principles of which were accepted by the Australian Government. The (wcs) document describes global problems to which Australians contribute and outlines solutions in which Australians should play a role. It also recommends the preparation of national strategies.

7.   The Strategy has been prepared jointly by government, industry, conservation and other groups following widespread public consultation.

8.   The main steps in formulating the Strategy were:

(a)   preparation of specially commissioned papers on aspects of living resource conservation;
(b)   discussion of the papers at a National Seminar in December 1981;
(c)   preparation of a discussion paper released in May 1982 for public comment;
(d)   consideration of over 550 written submissions on the discussion paper;
(e)   preparation of a Conference Draft Strategy and associated background papers; and
(f)   the convening of a National Conference in June 1983 at which a consensus was reached on the National Conservation Strategy for Australia now presented.

9.   The Strategy is submitted for consideration by governments and the wider community with a view to its adoption throughout Australia.

10.   Consideration of the Strategy and its implementation will take place within Australia's federal, constitutional, legislative and administrative framework. In addition the implementation of NCSA recommendations should have regard for social, economic, cultural and other relevant goals.

11.   It is not possible for a broad Strategy such as this to refer specifically to all conservation practices currently in place in Australia. In recent years much has been done in the area of living resource conservation. This document identifies broad strategic measures necessary to bring about properly integrated conservation and development practices in Australia; some of these measures are already being implemented.

12.   A National Conservation Strategy cannot be a static thing. It must be dynamic and should be reviewed and modified from time to time to take into account changing circumstances, new knowledge and emerging community attitudes.

Why a national conservation strategy

13.   Within Australia many agencies and organisations have been established to address particular aspects of living resource conservation and development and much has been achieved. Nevertheless, much remains to be done as a matter of urgency because:

(a)   the growth in Australia's population and the high standard of living enjoyed by most Australians place increasing demands on resources. An additional demand arises because Australia supplies commodities to other countries in return for imports of goods and services;
(b)   soil erosion, largely the result of inappropriate use and management of land, degrades dams and waterways, affects terrestrial and aquatic ecosystems and community facilities, and poses a major threat to agriculture and the Australian economy;
(c)   Australia is the driest inhabited continent and the quantity, quality and location of water

Michaelmas Cay on the Great Barrier Reef near Cairns. Tourists and seabirds live in harmony on this small island. Human presence has protected the nesting colonies from vandals and today there are always some birds nesting at any time of the year. (See also pp. 128–30)

The Daintree and its rainforest in the wet tropics of north-east Queensland is nominated for World Heritage listing. These rainforests lie along the coastal belt between Townsville and Cooktown. (See also pp. 128–30)

Jarrah forest. Strip mining for bauxite in south-western Australia has destroyed huge areas of the original forest which is being replanted to a variety of species. Yet jarrah is one of the world's most famous hardwoods. (See also p. 113)

influences how and where development may occur. Some river systems, wetlands and underground water resources are severely degraded. Water and soil salinity are now widespread problems;

(d) Australia has lost much of its native vegetation particularly through clearing for agriculture, grazing and urban development. In some areas this has contributed to soil salinity. Certain types of native vegetation communities are now scarce and harvesting rates of some of these communities are above sustainable levels;

(e) many native Australian plants and animals are endangered and some species have become extinct. Others have increased in numbers to nuisance levels. The habitats of many species have either been destroyed or are severely affected by human activities and by introduced animals and plants;

(f) some fisheries stocks are declining because of overharvesting and changes to the aquatic environment;

(g) estuarine and coastal environments which are highly productive nurseries for aquatic organisms are limited in Australia. Many sites are being either contaminated or reduced in area as a result of industrial, agricultural and urban pressures;

(h) significant sections of the coastal lands have been extensively modified while other areas are threatened by further development;

(i) there is an increasing threat to Australian living resources from introduced diseases and pests.

The strategy

Elements

14.   The essential elements of the National Conservation Strategy for Australia are:

(a) the **Definitions** of development and living resource conservation;

(b) the **Objectives** of living resource conservation;

(c) The **Principles**: namely integrating conservation and development, retaining options for future use, focusing on causes as well as symptoms, accumulating knowledge for future application, and educating the community; and

(d) a set of **Priority National Requirements and Actions**.

Definitions

15.   The Strategy recognises that living resource conservation and sustainable development are interdependent. This interdependence is emphasised by the definitions of **conservation** and **development** explained in detail in the World Conservation Strategy. These definitions have been adopted for the National Conservation Strategy for Australia:

(a) "**Conservation is the management of human use of the biosphere so that it may yield the greatest sustainable benefit to present generations while maintaining its potential to meet the needs and aspirations of future generations.** Thus conservation is positive, embracing preservation, maintenance, sustainable utilisation, restoration, and enhancement of the natural environment. Living resource conservation is specifically concerned with plants, animals and micro-organisms, and with those non-living elements of the environment on which they depend. Living resources have two important properties the combination of which distinguishes them from non-living resources: they are renewable if conserved; and they are destructible if not." (wcs).

(b)   **"Development is the modification of the biosphere and the application of human, financial, living and non-living resources to satisfy human needs and improve the quality of
human life.** For development to be sustainable it must take account of social and ecological
factors, as well as economic ones; of the living and non-living resource base; and of the long
term as well as the short term advantages and disadvantages of alternative actions." (wcs).

Actions which result in little or no modification of the biosphere can also satisfy human needs and
improve the quality of human life.

16.   Conservation and development are fundamentally linked by their dependence on living
resources. Both conservation and sustainable development require an attitude of stewardship,
especially towards those plants, animals and micro-organisms and the non-living resources on which
they depend, that could be destroyed if only short term human interests are pursued. To provide for
today's needs as well as to conserve the stock of living resources for tomorrow, both conservation and
development are necessary.

Objectives

17.   The three main objectives of living resource conservation identified in the World Conservation
Strategy have been adopted for the ncsa. They are:

(a)   **"to maintain essential ecological processes and life-support systems** (such as soil regeneration and protection, the recycling of nutrients, and the cleansing of waters), on which human
survival and development depend;
(b)   **to preserve genetic diversity** (the range of genetic material found in the world's organisms),
on which depend the breeding programmes necessary for the protection and improvement of
cultivated plants and domesticated animals, as well as much scientific advance, technical
innovation, and the security of the many industries that use living resources;
(c)   **to ensure the sustainable utilisation of species and ecosystems** (notably fish and other
wildlife, forests and grazing land), which support millions of rural communities as well as
major industries."

18.   An additional and no less important objective for Australia is **to maintain and enhance
environmental qualities** which make the earth a pleasant place to live in and which meet aesthetic
and recreational needs.

19.   In the context of these objectives, the role of development is to use resources to:

(a)   provide for the essential needs of individuals and society;
(b)   generate economic wealth which enables the community to enhance its standard of living and
to pursue educational, cultural and recreational interests including preserving its heritage; and
(c)   provide economic capacity which helps society to practise resource conservation which in turn
enables sustainable development.

20.   It follows that implementation of the Strategy must have regard for the general economic
climate, which has an important bearing upon the speed with which the Strategy can be implemented,
and for the inability of Australia to isolate itself from the world economic system. It also requires a
proper accounting of the costs and benefits to society.

21.   Australia's important role as a reliable supplier of food and resources is also relevant. Consistent with other objectives this can afford opportunities to minimise the extent of environmental degradation around the world. Trade also provides for the distribution of other economic benefits which themselves can facilitate global strategies for sustainable development.

Factors affecting achievements of objectives

22.   Some factors which may be obstacles to the achievement of NCSA objectives are:

(a)   the belief that Australia is so vast that it has an unlimited capacity to supply resources;

(b)   a lack of recognition that there is a limit to the ability of Australia's life support systems to withstand human impacts;

(c)   lack of recognition that, in the long term, continuous growth in population and per capita demand is inconsistent with the maintenance of essential ecological processes and life support systems;

(d)   lack of consistent application of guidelines designed to ensure that development is undertaken with minimum adverse environmental impact and that existing adverse impacts are ameliorated;

(e)   a lack of recognition that conservation applies to land and other resource uses and activities other than national parks and wildlife protection;

f)   a restriction of wider, more beneficial uses of living resources by some organisations with narrow, single-purpose aims;

g)   inadequate planning for the integration of conservation and development for a sustainable future;

(h)   insufficient co-ordination between the various bodies involved in making decisions about living resources;

(i)   uncertainty about which of local, state and federal levels of decision-making is appropriate for particular matters;

(j)   lack of understanding of and attention to the cultural, economic, legal, political and social factors which influence decisions about living resources;

(k)   fear of unemployment arising from implementation of conservation measures;

(l)   the higher visibility of the costs of many conservation measures as opposed to their benefits;

(m)   insufficient knowledge and understanding of the issues;

(n)   inadequate understanding among the general community of the nature and importance of living resource conservation and a lack of appropriately trained personnel; and

(o)   increasing pressures to develop granted Aboriginal lands that often have relatively few options for sustained development.

23.   **Some factors which may assist** in the achievement of NCSA objectives are:

(a)   a widespread appreciation of nature and increasing community awareness of the need for living resource conservation;

(b)   the existence of legislation for, and government agencies concerned with, the management of living resources;

(c)   the fact that resource users are increasingly willing to develop and accept voluntary guidelines for their operations;

(d)   increased opportunities for training in living resource management;

(e)   improvements in technology which allow adverse consequences of development to be avoided, minimised or remedied;

(f)   the increase in recent years of land use planning procedures which take account of public values;

(g)   the growth in the voluntary conservation movement, in professionalism in the public services and in awareness on the part of industry, which all assist long term community interests to be taken into account; and

(h)   decisions by Aboriginal landowners to agree to the use of granted lands as National Parks.

Strategic principles for achieving the objectives

24.   Achieving the ncsa objectives would be greatly assisted if the Australian community accepted the following strategic principles.

(a)   **Integrate conservation and development** and emphasise their interdependence and common ground. This requires:

    (i)   wider acceptance of the interrelated definitions of conservation and development adopted in this Strategy;

    (ii)   better understanding and acceptance of the relationship between development and conservation;

    (iii)   improved decision-making, which may involve changes to institutional and organisational arrangements, and more and earlier consultation between interested parties to minimise conflict;

    (iv)   a clear awareness of cost savings that may arise if development planning takes place within the framework of the Strategy. This could reduce delays and both short and long term costs;

    (v)   recognition that conservation is relevant to all areas of living resources and is not confined to a narrow range of activities;

    (vi)   increased appreciation of the effects of one land use on another and of the value of multidisciplinary planning; and

    (vii)   greater weight to be given to long term considerations.

(b)   **Retain options for future use**. This is necessary because:

    (i)   capacity to anticipate the needs and aspirations of future generations is limited.

    (ii)   information on ecosystems and their ability to absorb impacts is frequently insufficient for resource managers to predict the future effects of their use and technology cannot be relied upon to correct adverse effects; and

    (iii)   management flexibility is required to respond to changing circumstances.

(c)   **Focus on causes as well as symptoms**. This is necessary because:

    (i)   it is usually more efficient to anticipate problems and to take positive preventative action than merely to react to problems when they arise; and

    (ii)   problems with immediate economic or social costs attract most attention while chronic environmental problems may be overlooked, although their effects may be very significant in the longer term.

(d)   **Accumulate knowledge for future application.** This is necessary because:

    (i)   resource use generates knowledge and experience which can be reapplied elsewhere;

    (ii)   development and conservation decisions can be reassessed periodically in the light of new knowledge to yield increased benefits for the community;

    (iii)   the availability of new knowledge can lead to improved decision-making; and

    (iv)   risks are associated with conservation and development proposals which need to be assessed and compared with the short and long term benefits to the community.

(e)   **Educate the community** on the interdependence of sustainable development and conservation.

Priority national requirements

25.   Priority National Requirements identify the major goals of the Strategy and ways and means of implementing it.

Major goals

(a)   Ensure that living resource development is such as to optimise the quality of life of Australians.

(b)   Ensure that land which is suitable for many sustainable uses is used in a manner which retains, as far as possible, the greatest number of options for furthur use.

(c)   Ensure that land management practices are consistent with long term productivity of living resources.

(d)   Restore degraded and eroded lands.

(e)   Ensure that productive agricultural and forestry systems are used on a sustainable basis.

(f)   Conserve Australia's ground and surface water resources; and restore and maintain water quality.

(g)   Avoid further increases in and, where possible, reduce salinity of soils and water caused by human activities.

(h)   Conserve Australia's soil, with the aim of ensuring that soil is not lost or degraded as a consequence of any land management practice.

(i)   Preserve the genetic diversity of Australia's plant and animal species and ecosystems and of those introduced species which support plant and animal based industries.

(j)   Manage the impact of development on the coastline, on aquatic resources, on the quality of coastal waters and on critical habitats such as wetlands, estuaries, bays and reefs so that their ability to meet conservation and development objectives is not diminished.

(k)   Ensure that the increasing use of the aquatic enviroment is managed so that its ecological integrity is retained and its utility and productive capacity are sustained.

(l)   Ensure that pollution and its effects on Australia's living resources are minimised.

Priority national actions

26.   Priority National Actions identify more specific measures for achieving the objectives of the NCSA. In implementing the actions governments and non-government bodies will be able to:

(a)   use the large body of existing legislation, administrative arrangements and skills available at all levels of government and, where necessary seek the harmonisation of these skills through consultative processes; and

(b)   build on the knowledge, expertise, public awareness and participatory processes available within the community.

Improving the capacity to manage

27.   Education and training

(a)   Develop and support informal education and information programmes, including those conducted by voluntary and other non-government organisations, which promote throughout the community an awareness of the inter-relationships between the elements of the life support systems and which encourage the practice of living resource conservation for sustainable development.

(b)   Review, strengthen and develop in schools environmental education programmes which have regard for the basic objectives and principles of the NCSA.

(c)   Review, strengthen and develop training, retraining and extension programmes for professionals, technicians and users involved in planning and management of activities which impinge upon living resources, which have regard for the basic objectives and principles of the NCSA.

28.   Policy planning and co-ordination

(a)   Strengthen co-ordination of action in and co-operation between the Commonwealth and the States and among the States on living resource issues of national significance.

(b)   Establish machinery to improve communication and to promote co-operation between community groups, industry and governments on matters related to implementation of the NCSA.

(c)   Integrate land use planning and environmental assessment by encouraging a multidisciplinary approach (including socio-economic effects) to ensure that conservation and development issues are not addressed in isolation.

(d)   Conduct thorough environmental and socio-economic assessments of proposals and policies that are likely to have a significant effect on living resources.

(e)   Develop methods whereby the costs and benefits of conservation and development proposals can be quantified in order to assess their potential impact on society.

(f)   Take into account the cumulative effects of conservation and development proposals and policies at local, regional and national levels.

(g)   Review and where appropriate revise the charters of single purpose government authorities to enable them to take account of both conservation and development objectives.

(h)   Ensure that financial and other incentives promote sustainable development, and remove incentives or apply disincentives for inappropriate activities, subject to the careful assessment of the consequences for other national objectives.

29.   Legislation and regulations

(a)   Through a process of consultation with conservation, industry and other interested groups, further develop and publish environmental standards, codes of practice and guidelines for the purpose of better implementing the goals as set out in the NCSA. In certain circumstances it may be appropriate for industry to develop and publish voluntary codes of practice.

(b)   Through a process of consultation between governments, work towards the harmonisation within Australia of conservation and environment protection legislation having regard for the goals of the NCSA and the need to avoid unnecessary delays and duplication created by differing procedures.

(c)   Incorporate a statement of relevant conservation and development objectives in legislation involving the environment.

(d)   Encourage the provision of explanatory information to assist in the understanding of environmental legislation.

(e)   Encourage governments to examine existing legislation which may promote activities inconsistent with the NCSA having regard for other relevant priorities.

30.   Research

(a)   Strengthen research efforts to improve knowledge of the different life support systems, of their capability for being used for different purposes and of the management required to sustain their capability for those uses.

(b) Improve national co-ordination of environmental research so that effort is better directed toward agreed priorities, duplication is avoided and adequate communication exists between research agencies.

(c) Assess the ecological effects of introducing plants, animals and micro-organisms into Australia to provide a basis for strengthening quarantine procedures.

(d) Improve taxonomic and ecological knowledge of plant and animal species and their distribution, impacts and interrelationships.

31. International

(a) Strengthen consultative arrangements and information exchange between the Commonwealth and other government and non-government bodies concerning Australia's participation in international conservation agreements and programmes.

(b) Ensure that the objectives arising out of the NCSA and the WCS are taken fully into account in Australia's dealings with other countries.

(c) Promote international understanding of the importance of the unique physical character and ecology of Antarctica and seek appropriate forms of management for the continent.

Managing for sustainable yield while protecting life support systems

32. Reserves and Habitat Protection

(a) Assess and, where necessary, expand the conservation reserve system to ensure the comprehensive representation of ecosystems, species and genetic diversity of species and the protection of a range of reserves serving recreation, heritage and amenity needs including wilderness areas. Give priority to the inclusion of areas of exceptional diversity and to the isolated surviving remnants of past distributions, that is, relics and refugia.

(b) Retain and manage representative samples of natural landscapes and habitats in developed areas.

(c) Ensure parks and reserves are large enough to conserve species of flora and fauna under adverse conditions and in the longer term.

(d) Identify and manage habitats of economically and culturally important species.

(e) Promote the retention of native vegetation on all lands, including those used for agriculture, pastoralism, forestry, mining and transportation.

(f) Ensure existing tourism, recreation and other facilities associated with areas of high conservation value do not adversely affect this value. Encourage the establishment of new facilities outside areas of high conservation value.

(g) Maintain self-perpetuating wild populations of native plants and animals over as wide a range as possible.

(h) Assess the ecological and economic impact of the planned and accidental introduction of plants and animals and micro-organisms to Australia. Determine the acceptance, control or eradication of such entities as necessary.

(i) Control animal or plant pests in natural and managed ecosystems.

(j) Review statutory requirements such as those relating to fire and vermin control to promote consistency with overall land management priorities.

(k) Encourage the development of appropriate fire protection and wildfire management plans for areas of natural vegetation giving due consideration to the possible impact of uncontrolled fires on the community at large.

(l)   Use proven hygiene practices to protect sites of high conservation value, for example in relation to dieback.

(m)   Undertake and publish a national inventory of wetlands and flood plains and draw up a set of criteria for evaluating their conservation.

33.   Controlling pollution, wastes and hazardous materials

(a)   Evaluate more fully the effects of hazardous materials, effluents and wastes released into the environment or used by society, with a view to the more appropriate control, treatment and disposal of such materials.

(b)   Prevent the disposal of inadequately treated effluent and wastes on land and wetlands, in water courses and at sea and, by the application of the best practicable means, ensure that wastes and effluent are adequately treated and are disposed of appropriately. In the case of intractable wastes ensure that they are safely and permanently contained.

(c)   Support the "polluter pays principle" as defined by the OECD, which advocates that people should pay for the full cost of their actions and for the resources they use. Qualify the application of the principle where necessary to take into account changes over time in community and pollution standards and practices, and to ensure equity.

(d)   Minimise the release of pollutants into the atmosphere and conform to established air pollution standards.

(e)   Minimise the effects of wastes and pollution upon living resources by promoting the most efficient methods of extraction and use, by promoting the development of wastes as potential resources and by promoting recycling of resources.

34.   Using Living Resources

(a)   Encourage governments to develop and review from time to time overall co-ordinated policies for the use and management of Australia's living resources (including forests, flora, fauna, fisheries, soil and water).

(b)   Ensure that harvested stocks of terrestrial and aquatic living resources are sustainable and make public the basis on which harvesting rates are determined.

(c)   Ensure, in the harvesting of indigenous plants and animals, that the condition of the habitat is a factor in setting harvesting rates.

(d)   Recognise that sustainable use of native rather than introduced species may be practicable and desirable in some areas because of cultural and environmental values; and that species should not be introduced to Australia without due regard to the possible alternative use of native equivalents.

(e)   Develop and encourage the use of fisheries techniques that minimise incidental take.

(f)   Ensure, in the management of forest areas for timber production, that due regard is had for the various benefits and uses that the forest may provide and ensure that any benefits are sustainable.

(g)   Retain the viability of the native plant and animal communities by suitable management practices in those forest areas used for timber production.

(h)   Promote the substitution of plantation products for native forest products, recognising that plantation forestry can reduce demand on native forests.

(i)   Conserve viable native forests by avoiding clearing such forests for plantations and by concentrating plantations on previously cleared, or substantially cleared, areas.

(j)   Give high priority to the protection of rainforests and urgent consideration to those most threatened.

35.   Conserving Soils and Water

(a)   Maintain and where appropriate improve soil fertility, structure and productivity through improved land management practices.

(b)   Develop and encourage a wide range of measures including the retention of native vegetation and regeneration and, where necessary, the planting of vegetation to reduce the spread and ameliorate the effects of soil and water salinity.

(c)   Give high priority through a national soil conservation programme to intensifying soil conservation and restoration efforts to the level necessary to ensure sustainable agricultural and pastoral production and adequate protection of catchment areas, conservation areas, public lands, community facilities and marine and fresh water systems.

(d)   Recognise that water is a limited resource in Australia; promote its wise and efficient use and protect and where necessary improve its quality.

(e)   Take an integrated whole catchment approach to the management of water and related land resources.

(f)   Manage groundwater resources on a sustainable basis in respect of both quantity and quality.

(g)   Plan and manage water impoundments, flood mitigation and other water course works to take account of the requirements of fish and other wildlife and wetlands, estuarine, riverine forest and other ecosystems and affected aquifers.

(h)   Encourage the establishment of vegetation on degraded lands and avoid further clearing where the land use cannot be sustained.

(i)   Establish a suitable administrative framework for the review of arid land use and management.

The Future

36.   The success of the NCSA will depend on continuing government and public support and on establishing ways of evaluating its implementation, on ensuring that it continues to be updated and on involving the Australian community in its further development and application.

Appendix 3

Official List of Australian Endangered Vertebrate Fauna

(Dates shown in parentheses represent the date of the last authentic record.)

	Scientific name	*Common name*
Mammals	*Bettongia lesueur*	burrowing bettong (boodie)
	Bettongia penicillata	brush-tailed bettong (woylie)
	Bettongia tropica	northern bettong
	Caloprymnus campestris (1935)	desert rat-kangaroo
	Lagorchestes asomatus (1931)	central hare-wallaby
	Lagorchestes hirsutus	rufous hare-wallaby
	Lagorchestes leporides (1890)	eastern hare-wallaby
	Lagostrophus fasciatus	banded hare-wallaby
	Macropus greyi (1927)	toolache wallaby
	Onychogalea fraenata	bridled nailtail wallaby
	Onychogalea lunata (1930)	crescent nailtail wallaby
	Petrogale sp. nov.	Proserpine rock-wallaby
	Potorous platyops (c. 1875)	broad-faced potoroo
	Gymnobelideus leadbeateri	Leadbeater's possum
	Lasiorhinus krefftii	northern hairy-nosed wombat
	Chaeropus ecaudatus (1926)	pig-footed bandicoot
	Perameles bougainville	western barred bandicoot
	Perameles eremiana (1931)	desert bandicoot
	Macrotis lagotis	greater bilby (dalgyte)
	Macrotis leucura	lesser bilby
	Thylacinus cynocephalus (1936)	thylacine (Tasmanian tiger)
	Myrmecobius fasciatus	numbat
	Sminthopsis longicaudata	long-tailed dunnart
	Sminthopsis psammophila	sandhill dunnart
	Conilurus albipes (1875)	rabbit eared tree-rat
	Leporillus apicalis (c. 1840)	lesser stick-nest rat
	Leporillus conditor	greater stick-nest rat
	Notomys amplus (1896)	short-tailed hopping-mouse
	Notomys aquilo	northern hopping-mouse
	Notomys fuscus	dusky hopping-mouse
	Notomys longicaudatus (1906)	long-tailed hopping-mouse
	Notomys macrotis (c. 1840)	big-eared hopping-mouse
	Notomys mordax (1840)	Darling Downs hopping- mouse
	Pseudomys fieldi (1896)	Alice Springs mouse
	Pseudomys praeconis	Shark Bay mouse
	Xeromys myoides	false water-rat
	Zyzomys pedunculatus	central rock-rat

Scientific name	Common name
Balaenoptera musculus	blue whale
Megaptera novaeangliae	humpback whale
Eubalaena australis	southern right whale

Birds

Scientific name	Common name
Pterodroma leucoptera	Gould's petrel
Pterodroma solandri	providence petrel
Puffinus carneipes hullianus	Lord Howe Island flesh-footed shearwater
Sula abbotti	Abbott's booby
Fregata andrewsi	Christmas Island frigatebird
Accipiter radiatus	red goshawk
Pedionomus torquatus	plains-wanderer
Tricholimnas sylvestris	Lord Howe Island woodhen
Anous tenuirostris	lesser noddy
Cyanoramphus novaezelandiae cookii	Norfolk Island parrot
Cyclopsitta diopthalma coxeni	Coxen's fig parrot
Geopsittacus occidentalis	night parrot
Neophema chrysogaster	orange-bellied parrot
Pezoporus wallicus	ground parrot
Polytelis alexandrae	Alexandra's parrot
Psephotus chrysopterygius	golden-shouldered parrot
Psephotus dissimilis	hooded parrot
Psephotus pulcherrimus (1922)	paradise parrot
Ninox novaeseelandiae royana	Norfolk Island boobook owl
Ninox squamipila natalis	Christmas Island hawk-owl
Podargus ocellatus plumiferus	marbled frogmouth
Atrichornis clamosus	noisy scrub bird
Malurus coronatus	purple-crowned fairy-wren
Amytornis dorotheae	Carpentarian grasswren
Amytornis textilis textilis	thick-billed grasswren
Dasyornis longirostris	western bristlebird
Dasyornis brachypterus	eastern bristlebird
Dasyornis broadbenti littoralis (1940)	rufous bristlebird
Strepera graculina crissalis	Lord Howe Island currawong
Psophodes nigrogularis	western whipbird
Pardalotus quadragintus	forty-spotted pardalote
Zosterops albogularis	Norfolk Island silvereye
Lichenostomus melanops cassidix	helmeted honeyeater
Manorina flavigula melanotis	black-eared miner
Drymodes superciliaris colcloughi (1915)	northern scrub-robin

Frogs

Scientific name	Common name
Arenophryne rotunda	
Philoria frosti	Mt Baw Baw frog

	Scientific name	Common name
	Rheobatrachus silus	platypus frog
	Cophixalus concinnus	elegant microhylid
	Cophixalus saxatilis	rock-dwelling microhylid
	Litoria longirostris	long-nosed tree frog
Reptiles	*Dermochelys coriacea*	leathery turtle
	Pseudemydura umbrina	western short-necked swamp tortoise
	Lerista lineata	lined burrowing skink
	Ctenotus lancelini	Lancelin Island striped skink
	Pseudemoia palfreymani	Pedra Branca skink
	Ophidiocephalus taeniatus	bronzebacked legless lizard
	Aprasia parapulchella	
	Neelaps calonotus	black-striped snake
	Hoplocephalus bungaroides	broad-headed snake
Fish	*Prototroctes maraena*	Australian grayling
	Macquaria australasica	Macquarie perch
	Maccullochella macquariensis	trout cod

Research or rot?

There are a number of ways to strengthen environmental research in Australia, and we need to assess them carefully. These include:

Developing a national strategy for environmental research, linked to the national conservation strategy and the federal government's recent statement on the environment.

Establishing a national environmental research and development corporation to plan and fund research not covered by existing research corporations.

Conducting more long-term, large-scale and multidisciplinary research programmes, extending inventories and undertaking more monitoring of Australian natural resources and ecosystems, which is necessary for their effective management.

Improving recruitment and training of environmental research scientists through better links between universities, the CSIRO and other government agencies.

Enhancing the interaction and flow of information.

Increasing funds for environmental research.

Introducing a levy on relevant industries, such as forestry, tourism and energy, to help fund environmental research, particularly where this research will contribute to sustainable or more efficient use of resources.

Providing funds for education programmes.

These matters are far removed from the issues that dominate political and public debate about the invironment. But in the long term, they will be more important in determining how successful we are in tackling the environmental problems we face.

This article is adapted from CSIRO analyst Richard Eckersley's submission to the Australian Science and Technology Council's current review of the state of environmental research in Australia.

(*The Australian Magazine*, 1–2 December 1989)

Further reading

Allen, Robert, IUCN, UNEP, WWF, *How to Save the World: Strategy for World Conservation,* Kogan Page, London, 1980.

Atlas of Australian Resources, Australian Government Publishing Service, Canberra, 1987.

Department of the Arts, Heritage and the Environment, *State of the Environment in Australia 1985,* Australian Government Publishing Service, Canberra, 1985.

IUCN, UNEP, WWF, *World Conservation Strategy,* 1980.

Leigh, J., Boden, R., Briggs, J., *Extinct and Endangered Plants of Australia,* Macmillan, Australia, 1984.

Morgan, D. G., ed., *Biological Science: The Web of Life,* Australian Academy of Science, Canberra, 1981.

Myers, Norman, ed., *The GAIA Atlas of Planet Management,* Pan Books, London and Sydney, 1985.

Ovington, Derrick, *Australian Endangered Species,* Cassell, Australia, 1978.

Serventy, Vincent, *Australia's World Heritage Sites,* MacMillan, Australia, 1986.

Teacher's Guide to National Conservation Strategy, Australian Government Publishing Service, Canberra, 1987.

Whitelock, Derek, *Conquest to Conservation,* Wakefield Press, Adelaide, 1985.

There are dozens of other books that deal with the topic. State Environment Centres, National Parks and Wildlife Service bookshops, and museum, zoo and botanic garden bookshops all offer a wide range.

Index